# Operating at Peak Efficiency
## A Technician's Guide to Servicing Hvac/r Equipment

Billy C. Langley

Business News
Publishing Company
Troy, Michigan

Copyright © 1995
Business News Publishing Company

All rights reserved. Except as permitted under the United States Copyright Act of 1976, no part of this publication may be reproduced or distributed in any form or means, or stored in a database or retrieval system, without the prior written permission of the publisher, Business News Publishing Company.

---

**Library of Congress Cataloging in Publication Data**

Langley, Billy C., 1931-
    Operating at peak efficiency: a technician's guide to servicing hvac/r equipment / Billy C. Langley
        p. cm.
    ISBN 1-885863-07-1
    1. Heating--Equipment and supplies--Maintenance and repair. 2. Ventilation--Equipment and supplies--Maintenance and repair. 3. Air Conditioning--Equipment and supplies--Maintenance and repair.  I. Title.
TH7015.L36           1995           95-15991
697'.0028'8--dc20                   CIP

---

Editors:  Joanna Turpin, Carolyn Thompson
Art Director:  Mark Leibold

This book was written as a general guide. The author and publisher have neither liability nor can they be responsible to any person or entity for any misunderstanding, misuse, or misapplication that would cause loss or damage of any kind, including loss of rights, material, or personal injury, or alleged to be caused directly or indirectly by the information contained in this book.

Printed in the United States of America
7 6 5 4 3 2 1

# Foreword

This manual was written to provide the service technician with the procedures necessary to bring heating, air conditioning, and refrigeration systems, including heat pumps, to full operating efficiency. This manual was not intended to present "standard" service procedures but rather to provide advanced information and procedures that, when followed, cause the equipment to operate as it was designed by the manufacturer.

When used properly, the procedures presented in this manual will ensure equipment operates more economically and to full capacity and has a longer life with a minimum amount of repairs.

# Acknowledgments

It would have been impossible to produce this manual without the help of many manufacturers and friends who are concerned with this industry. Their contributions over the past 35 years have made a manual of this type possible. They have been most helpful in providing information that, in part, made this text feasible. My sincere appreciation goes to all those companies and individuals who value this industry as much as I do.

Billy C. Langley

# Table of Contents

Chapter 1
Why Fine Tune Equipment? ............................................. 1

Chapter 2
Electric Heating ............................................................ 5

Chapter 3
Gas Heating (Natural and LP) ....................................... 17

Chapter 4
Oil Burners ................................................................. 31

Chapter 5
Air Conditioning Systems and Heat Pumps
(Cooling Mode) ........................................................... 41

Chapter 6
Heat Pumps (Heating Mode) ......................................... 55

Chapter 7
Refrigeration .............................................................. 65

Chapter 8
Megohmmeters ........................................................... 75

Chapter 9
Pressure-Enthalpy Diagrams ......................................... 79

Chapter 10
Estimating Annual Operating Costs .............................. 113

Chapter 11
Estimating Annual Heating Requirements .................... 123

Appendix A
Air Conditioning Formulas (Non-Psychrometric) ........... 131

Appendix B
Worksheets ............................................................... 133

# Chapter 1

# Why Fine Tune Equipment?

There are many reasons for fine tuning heating, air conditioning, and refrigeration equipment, such as energy conservation, cost effectiveness, less need for new power plants, improved performance, and equipment longevity. It is true it takes more time to fine tune equipment so that it operates at peak efficiency, and this comes at a greater cost to the customer. However, when the benefits are properly explained, most are willing to pay the service fee, which is usually saved many times over through the more efficient and economical operation of the equipment. Also, the technician who has the ability and desire to fine tune equipment can charge extra for his labor and will always be in demand by the public.

## ENERGY CONSERVATION

Because of the constantly decreasing sources of oil, every possible step must be made to conserve its use. Fine tuning heating, air conditioning, and refrigeration equipment is a very good place to start, because air conditioning units are the largest users of electricity in residential and many commercial buildings. When heating, air conditioning, and refrigeration units are operating at peak efficiency, this percentage of power consumption is reduced by sometimes as much as 25%. Several tests have proven that nine out of ten heating, air conditioning, and refrigeration units are operating at a reduced efficiency that is between 10% and 40% of their rating.

## COST EFFECTIVENESS

When heating, air conditioning, and refrigeration equipment is operating properly, less energy is used and the cost of operation is reduced. Thus, the customer is saving money. When the technician fine tunes the equipment so it operates more economically, the customer is more satisfied and is much more pleased with the service provided. The service technician can usually charge more for providing this fine tuning service, because the customer is more willing to pay for good, thorough service that ensures the equipment is operating as designed. Because of this, both parties are satisfied and the customer is more apt to recommend the technician to others who may want this type of service.

## Less Need for New Power Plants

When heating, air conditioning, and refrigeration equipment is working at peak efficiency, power generating plants can be operated at less than full capacity. The power company not only saves on operating expenses, but the need to build new power plants to furnish power to inefficiently operating equipment is no longer a consideration. Thus, the need for costly power plant construction and operation is eliminated.

## Improved Performance

Properly tuned heating, air conditioning, and refrigeration equipment performs better than equipment that is not properly adjusted. The customer will be more satisfied with the operation of fine tuned equipment. In addition, the unit will keep the building more comfortable, the process refrigeration and heating equipment will satisfy the demands much more easily, and there will be less service and maintenance required.

## Equipment Longevity

Fine tuned equipment has the proper amount of refrigerant and oil flowing through the system to maintain properly lubricated components at their desired operating temperatures. Properly lubricated equipment operating at the desired temperature usually lasts much longer than equipment that does not have these characteristics. Thus, the customer is saved the cost of having to replace the equipment, and major repairs are either eliminated or postponed to a much later date.

## Checking Pressures, Temperatures, and Airflow

A check of the operating refrigerant pressures and temperatures, along with determining the airflow through the unit and the temperature rise or drop of the air, is usually all that is needed to determine the efficiency of the unit. The proper adjustment of one or more of these operating factors usually determines the Btu output of the unit. Of course, the ductwork, insulation, and the condition of the structure will determine, to a great extent, whether or not the system will provide the desired conditions inside the building. All of these factors must be considered when fine tuning equipment to ensure more complete customer satisfaction. Remember, the best equipment will not perform if it is installed poorly. Installation is also a part of fine tuning a unit. The unit may be operating at peak efficiency, but if the conditioned air cannot reach the space or if it is not properly distributed, the result will still be poor operation. Give the installation a thorough inspection, and inform the customer of anything that can, or must, be done to obtain optimum efficiency and satisfaction.

# Chapter 2

# Electric Heating

The efficiency of any unit is the relationship of the amount of heat input to the amount of heat output. Electric heaters are considered to be 100% efficient. That is, for every Btu input, one Btu is delivered by the furnace. However, this is not completely true, because some energy is always lost when one form of energy is changed to another. But for our purposes, this statement is accurate enough. The purpose of fine tuning equipment is to change as much input heat as possible into heat for the building or process being treated. The system components must be clean and in good working condition, and clean air filters must be installed.

Electric heating units are used in both residential and commercial applications. They are available in both single- and three-phase element designs. Because of the lower discharge air temperatures, it is important that electric heating units operate as they were designed to prevent drafty conditions.

## Adjustment Process

Checking and adjusting the capacity of electric heating units is probably the most simple procedure for all of the various types of units available. The steps used to check the efficiency and capacity of electric heating units are discussed below. There is also a worksheet at the end of this chapter with the procedures outlined.

## Required Test Instruments

The proper test instruments are needed to make accurate measurements of the product being tested. The old procedure of merely looking at something or feeling a line no longer indicates what is happening with the unit.

There are many brands and models of test instruments available. Which particular instrument to use is the user's choice. At any rate, accurate test instruments must be used for testing the efficiency of electric heating units. The exact procedures for instrument use are found in the manufacturer's operating instructions. The instruments must be properly cared for and their accuracy maintained for optimum performance.

This chapter will discuss the specific procedures used to fine tune electric heating units, not specific instruments. The following is a list of the basic test instruments required for testing and adjusting the efficiency of electric heating units:

1. Dry bulb thermometer
2. Ammeter
3. Voltmeter (a wattmeter may be used in place of the ammeter and voltmeter)

### Dry Bulb Thermometer

A dry bulb (db) thermometer is used for checking air temperatures, Figure 2-1. Determining the temperature rise of the air as it passes through the heating unit is a very important step in the fine tuning process. The dry bulb temperature is a measure of the heat absorbed from the heating elements by the air. The air temperature is measured in two locations: the return air stream and the discharge air stream, Figure 2-2. Be sure to take the temperature readings <u>after</u> the mixing of any air and within 6 ft of the air handling unit. The thermometers must be placed where the radiant heat from the elements cannot be measured by the ther-mometer. Radiant heat can cause a faulty temperature reading and an incorrect test.

Two different thermometers that have been tested and found to produce exactly the same readings under the same set of circumstances should be used. If two thermometers that read exactly the same temperatures are not available, then use one thermometer to measure both

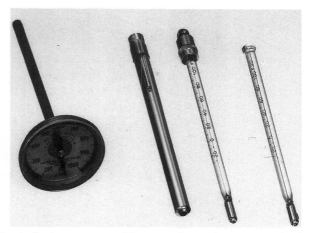

Figure 2-1. Dry bulb thermometers (Courtesy, Dwyer Instruments, Inc.)

# ELECTRIC HEATING

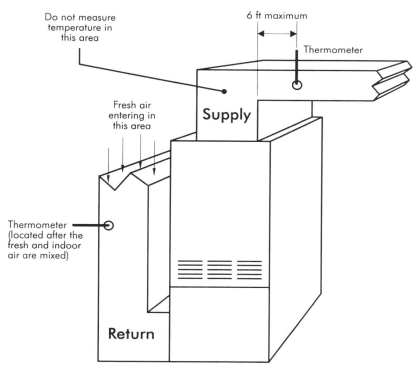

Figure 2-2. Measuring circulating air temperature

temperatures. An electronic thermometer that has separate, properly adjusted leads also produces the desired results. The best operating temperature rise of an electric heating unit is about 40° to 50°F. This allows the blower to run longer, distributing the air more evenly throughout the building.

## Ammeter

The ammeter is used to determine the amount of electrical current used by the unit. The clamp-on type ammeter is the most popular, because the amperage can be measured without separating the wire, Figure 2-3. These instruments are the most accurate when the wire being measured is in the center of the tongs.

## Voltmeter

The voltmeter is used to measure the voltage in a wire, Figure 2-4. The analog-type meter is the most accurate when the indicator is in the center of the scale. The leads should be checked regularly to ensure they are in good working condition. When the insulation becomes worn or cracked, the leads should be replaced. A firm, solid fit between the meter and the lead must be maintained. A loose fitting lead can give an improper voltage reading.

Figure 2-3. Clamp-on ammeter (Courtesy, TIF Instruments, Inc.)

Figure 2-4. Digital multimeters (Courtesy, A.W. Sperry Instruments, Inc.)

## Wattmeter

The wattmeter is used to measure the voltage to the unit and the total wattage used by the unit, Figure 2-5. These instruments are usually more accurate than the voltmeter and ammeter to determine the total wattage used by the unit. However, wattmeters are more expensive. The analog-type wattmeter is the most accurate when the indicator is reading at the midscale point.

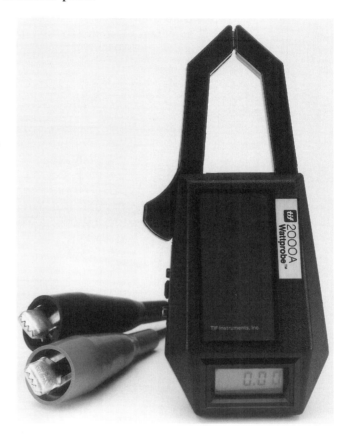

Figure 2-5. Clamp-on digital wattmeter (Courtesy, TIF Instruments, Inc.)

# KILOWATT INPUT

Depending on the size and type of installation, both single- and three-phase heating elements are used. When determining the kilowatt input, be sure to measure the total voltage and amperage to the unit, including the fan motor, because it also adds heat to the air. When more than one heating element is used, check the fan motor separately to prevent adding the motor amperage to each element. The measurements should be taken at a close disconnect switch or where the electricity enters the unit, not at a distant point.

## DETERMINING ELECTRIC HEATING UNIT CAPACITY

To determine the capacity of an electric heating unit, first determine the watt and Btu input to the unit by performing the following steps:

1. Measure the voltage and amperage to the fan motor and each heating element.
2. Add together the amperage draw of each heating element and the fan motor.
3. Determine the unit wattage by multiplying the voltage by the amperage. Use the following formula:

$$W = V \times I$$

where:  W = wattage
V = voltage
I = amperage

4. Determine the unit Btu output by multiplying the wattage by 3.413. Use the following formula:

$$Btu = W \times 3.413$$

5. Determine the blower cfm by dividing the Btu by 1.08 x ΔT. Use the appropriate formula:

$$cfm = \frac{Btu}{1.08 \times \Delta T}$$

$$cfm = \frac{kW \times 3413}{1.08 \times \Delta T}$$

$$cfm = \frac{W \times 3.413}{1.08 \times \Delta T}$$

where:  1.08 = specific heat of air constant
ΔT = temperature rise
cfm = cubic feet per minute
kW = kilowatts

Make certain that all the return and supply air grilles are open and unobstructed. If an air conditioning unit is included, make certain that the coils and filters are clean.

6. Determine the ΔT by using the following steps:
    a. Allow the unit to run continuously for about ten minutes.

# ELECTRIC HEATING

    b. To avoid thermometer error, use the same thermometer to measure the return and supply air temperatures.
    c. Measure the temperatures at a point where the thermometer cannot "see" the heat source (see Figure 2-2). True air temperature cannot be measured when radiant heat is sensed by the thermometer.
    d. The air temperature measurements must be taken within 6 ft of the air handling unit. The return air temperature can be taken at the return air grille if it is located close to the unit, as in Figure 2-2. The air temperature taken at the supply air grille is not usually accurate enough for this purpose.
    e. When more than one discharge or return air duct is connected to the plenum, use the average temperature (AT). For example, Duct 1 = 115°F, Duct 2 = 110°F, and Duct 3 = 108°F. Use the following formula:

$$AT = \frac{Duct\ 1 + Duct\ 2 + Duct\ 3}{Number\ of\ ducts}$$

$$AT = \frac{115°F + 110°F + 108°F}{3} = 111°F$$

    f. Be sure to take the temperature measurements <u>after</u> any source of mixed air, such as a fresh air intake.
7. Check the manufacturer's specifications for the particular unit to see if these conditions meet the design criteria.

When the cfm is too high, slow the blower. When the discharge cfm is too low, speed up the blower. When a direct drive motor is used, the cfm can be changed by moving the electric wire to a lower or higher speed terminal on the motor or in the furnace wiring junction box. Check the unit wiring diagram to make sure the correct connections are made. Otherwise, damage to the motor could occur.

When a direct drive motor cannot be wired to deliver the correct cfm, use the next higher speed and reduce the air inlet to the blower. Be sure to make the reduction on the blower end opposite the motor to prevent overheating of the motor. Adjust the baffle to deliver the correct cfm, and secure the baffle in place. The baffle can be made from a piece of sheet metal cut and screwed in place on the blower housing, Figure 2-6.

When a belt drive blower is used, the cfm can be changed by use of an adjustable pulley on the motor shaft. To increase the cfm, close the pulley halves. To reduce the cfm, open the pulley halves. *Note: ΔT = discharge air temperature minus return air temperature.*

When the heating unit is operating at its rated capacity but is not heating the building sufficiently, there are other options available. First, open all of the supply grilles, measure the cfm from each one, and total

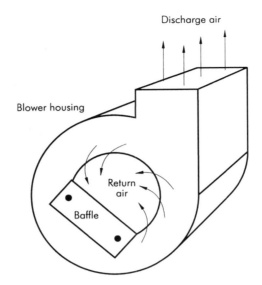

Figure 2-6. Blower inlet adjustment

them. If the cfm is 10% less than that determined in Steps 5 and 6 (listed previously), air is leaking from the duct system. Find the leak and repair it.

The next step is to measure the discharge air temperature as it leaves the supply air grille located farthest from the heating unit. The difference in discharge air temperature at the unit and this reading should not exceed more than 3° to 5°F. If the difference is more than 5°F, the duct needs more insulation.

When the duct system meets this criteria, the only other alternative is to add more capacity. This can be done by either adding more heat strips or installing some with higher capacity ratings. Table 2-1 lists the temperature rise, the kW rating of the furnace, and the corresponding cfm. This information can be used to reduce the time required to get a very close estimate of how efficient the unit is operating.

| Cfm | 5 kW | 7.5 kW | 10 kW | 12.5 kW | 15 kW | 17.5 kW | 20 kW |
|---|---|---|---|---|---|---|---|
| 300 | 53 | — | — | — | — | — | — |
| 400 | 59 | 79 | — | — | — | — | — |
| 500 | 32 | 47 | 63 | — | — | — | — |
| 600 | 26 | 39 | 53 | — | — | — | — |
| 700 | 23 | 34 | 45 | 57 | 68 | — | — |
| 800 | 20 | 30 | 40 | 49 | 59 | 69 | 79 |
| 900 | 18 | 26 | 35 | 44 | 53 | 61 | 70 |
| 1000 | 16 | 24 | 32 | 40 | 47 | 55 | 63 |
| 1100 | 14 | 22 | 29 | 36 | 43 | 50 | 57 |
| 1200 | 13 | 20 | 26 | 33 | 40 | 46 | 53 |
| 1300 | 12 | 18 | 24 | 30 | 37 | 43 | 49 |
| 1400 | 11 | 17 | 23 | 28 | 34 | 40 | 45 |
| 1500 | — | 16 | 21 | 26 | 32 | 37 | 42 |
| 1600 | — | — | 20 | 25 | 30 | 35 | 40 |
| 1700 | — | — | 19 | 23 | 28 | 32 | 37 |
| 1800 | — | — | 18 | 22 | 26 | 30 | 35 |
| 1900 | — | — | 17 | 21 | 25 | 29 | 33 |
| 2000 | — | — | 16 | 20 | 24 | 28 | 32 |

Table 2-1. Temperature rise in °F

OPERATING AT PEAK EFFICIENCY

# Electric Heating Unit Capacity Worksheet

Introduction: When a customer complains about insufficient heat from an electric heating unit, the service technician should, as a first step, determine if the heating elements are delivering the amount of heat they were designed to deliver. It could be that one or more of the heating elements is not operating properly. This is a fairly simple test and is easily performed.

Tools Needed: ammeter, voltmeter, accurate thermometer, and tool kit. An accurate wattmeter may be used in place of the ammeter and the voltmeter.

Procedures:

1. Turn off the electricity to any other equipment that is used in conjunction with the electric heating unit.

2. Set the thermostat to demand heat. Allow the heater to operate for approximately 10 minutes so the temperatures will stabilize.

3. Measure the voltage and amperage of each heating element and record:

|  | Volts | Amps |
|---|---|---|
| Motor: 1. | _____ | _____ |
| Heater: 1. | _____ | _____ |
| 2. | _____ | _____ |
| 3. | _____ | _____ |
| 4. | _____ | _____ |
| 5. | _____ | _____ |
| Sum: | _____ | _____ |

4. Determine the total wattage of the heat strips and record. Use the following formula:

$$W = V \times I$$

W = _____

5. Determine the Btu output of the heat strips and record. Use the following formula:

$$Btu = W \times 3.413$$

Btu = _____

Copyright © Business News Publishing Company

6. Determine the cfm of the blower and record. Use the following instructions:

    - Use the same thermometer, or two that measure exactly the same, to measure the return and supply air temperatures.

    - Do not measure the temperature in an area where the thermometer can sense the radiant heat from the heat strips (see Figure 2-2). True air temperature cannot be measured if the thermometer senses radiant heat.

    - Take the temperature measurements within 6 ft of the air handler. Measurements taken at the return and supply grilles that are at too great a distance from the unit are not usually accurate enough.

    - Use the average temperature when more than one duct is connected to the supply air plenum.

    - Be sure the air temperature has stabilized before taking the temperature measurements.

    - Take the temperature measurements downstream from any source of mixed air. Use the following formula:

    $$\text{cfm} = \frac{\text{Btu}}{1.08 \times \Delta T}$$

    cfm = _____

7. Is this what the manufacturer rates the equipment? _____

> Multiple tear-out copies
> of this worksheet
> can be found in
> Appendix B.

# Chapter 3

# Gas Heating (Natural and LP)

The purpose of adjusting the efficiency of any piece of equipment is to deliver the most energy possible to the conditioned space or process. The efficiency of a gas-fired furnace or boiler is the difference between the input and the actual amount of heat produced by the unit. The efficiency adjustment must provide the greatest gain without producing undesirable conditions. In mathematical terms, efficiency is explained as:

$$\text{Efficiency} = \frac{\text{Useful energy out}}{\text{Total energy}}$$

The components must be clean and in good working condition, and clean air filters must be installed.

## ADJUSTMENT PROCESS

There are several closely related factors involved in the adjustment process. When an adjustment is made to one of these factors, the other factors are also affected. There is a proper order to follow when making efficiency adjustments to gas-fired equipment. The general categories of these factors are as follows:

- Burner performance
- Heat exchanger operation
- Overall performance of the combustion process

### Burner Performance

The performance of the burner plays a very important part in overall unit efficiency. The burner is the place where gas and air are mixed in the proper proportions to gain maximum efficiency. Should something happen to upset this gas-air relationship, the unit's efficiency will drop in direct proportion. Therefore, the burner must be kept clean and properly adjusted.

### Heat Exchanger Operation

The heat exchanger is the component that transfers heat from the flue gases to the circulating air. It also provides a path for the escape of flue gases to the atmosphere. Should the flue passages become clogged with soot, lint, scale, or any other foreign material, the flue gases cannot flow through the passages properly. The air flowing to the burners will be reduced causing an inefficient and hazardous flame condition. Thus, two things occur. First, the desired amount of heat cannot be removed from the flue gases and a very hazardous condition exists. Second, when the flue gases cannot escape out of the unit properly, they will flow into the building causing the occupants to become ill and perhaps even die.

### Overall Performance of the Combustion Process

There are several factors that affect the overall combustion process and the unit's efficiency, such as the amount of combustion air to the unit, the amount of gas supplied to the burner, and the proper mixing of the gas and air.

## REQUIRED TEST INSTRUMENTS

The proper test instruments are needed to make accurate measurements of the product being tested. The old procedure of merely looking at a component or feeling a line no longer indicates what is happening with the unit.

There are many brands and models of test instruments available. Which particular instrument to use is the user's choice, but accurate test instruments must be used when fine tuning gas-fired equipment. The exact procedures for instrument use are found in the manufacturer's operating instructions. The instruments must be properly cared for and their accuracy maintained for optimum performance.

This chapter will discuss the specific procedures used to fine tune gas-fired heating units, not specific instruments. The following is a list of basic test instruments required for combustion testing and adjusting the efficiency of a gas-fired heating unit:

1. Gas manifold pressure gauge, or a U-tube manometer
2. Dry bulb thermometer
3. Flue gas temperature thermometer
4. Draft gauge
5. Carbon monoxide analyzer

6. Carbon dioxide analyzer with the appropriate combustion efficiency chart or slide rule to use in combination with the various test results to determine combustion efficiency
7. Velometer and other airflow measuring instruments

### Gas Manifold Pressure Gauge

The gas manifold pressure gauge is used to measure the gas pressure in the manifold pipe just before it passes through the orifices and into the burners. The gauge must be connected between the gas pressure regulator and the main burner orifices, Figure 3-1.

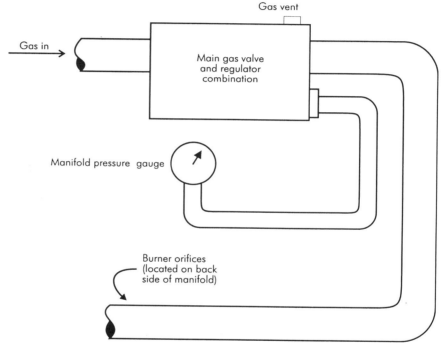

Figure 3-1. Gas manifold pressure gauge connection

Checking the gas manifold pressure gauge must be one of the first steps performed when fine tuning gas heating equipment, because if the heat is not getting in, there is no way to get it out.

The gauge may be a normal type pressure gauge, which measures in inches of water column (w.c.) or a U-tube manometer, which also measures the gas pressure in inches of w.c., Figure 3-2. Manufactured gas should have a manifold gas pressure of 2 to 3.5 inches w.c.; natural gas should have between 3 and 3.5 inches w.c.; and LP gas should be set at 11 inches w.c. Always check the unit nameplate to ensure the manufacturer's specifications are met.

Figure 3-2. U-tube manometer (Courtesy, Bacharach, Inc.)

### *Dry Bulb Thermometer*

A dry bulb (db) thermometer is a normal type thermometer used for checking air temperatures (see Figure 2-1). Determining the temperature rise of the air as it passes through the heating unit is a very important step in the fine tuning process. The dry bulb temperature is a measure of the heat absorbed from the heat exchanger by the air. The air temperature is measured in two locations: the return air stream and the discharge air stream, Figure 3-3. Be sure to take the temperature readings <u>after</u> the mixing of any air. The thermometer must be placed where the radiant heat from the elements cannot be measured by the thermometer. Radiant heat can cause a faulty temperature reading and an incorrect test.

Two different thermometers that have been tested and found to produce exactly the same readings under the same set of circumstances should be used. If two thermometers that read exactly the same temperatures are not available, then use one thermometer to measure both temperatures. An electronic thermometer that has separate, properly adjusted leads also produces the desired results. The best operating temperature rise of a gas-fired heating unit is about 70°F. This temperature allows the blower to run longer, distributing the air more evenly throughout the building.

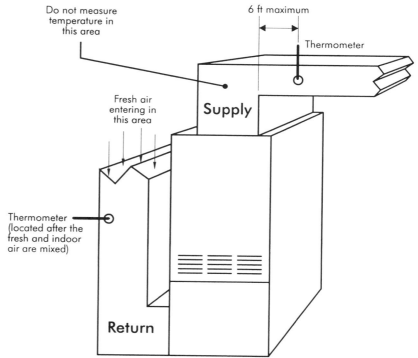

Figure 3-3. Measuring circulating air temperature

### Flue Gas Temperature Thermometer

This type of thermometer is used for measuring higher than normal temperatures. Generally, their range is about 200° to 1000°F, Figure 3-4. They are used to measure the temperature of the flue gases in the vent system, or the stack temperature. The flue gas temperature of a standard gas furnace or boiler should not be more than 480°F higher than the ambient or combustion air temperature. The minimum flue gas temperature on a standard gas furnace or boiler should be 350°F.

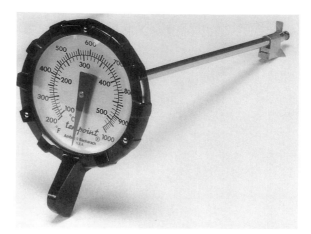

Figure 3-4. Flue gas temperature thermometer (Courtesy, Bacharach, Inc.)

If the flue gas temperature is too high, not enough heat is being removed from the burning gas. This may be due to low circulating airflow, a dirty heat exchanger, or the unit being over-fired. The causes of the problem must be found and corrected.

The flue gas temperature of a high efficiency gas furnace will range from about 100° to 125°F. Check the manufacturer's specifications for the specific temperature range. The other combustion factors will remain the same.

## Draft Gauge

The draft gauge is an instrument used for measuring small pressures, Figure 3-5. It is used to measure the rate at which flue products are removed from the unit heat exchanger. The reading is taken on the chimney side of the draft diverter, Figure 3-6.

A negative pressure reading indicates that there is sufficient air movement through the combustion zone to allow proper and complete combustion. If the reading is past the halfway point on the negative side of the scale, there is too much draft through the unit causing excess heat to be drawn out of the unit and inefficient operation. When the hand approaches zero or goes to the positive side of the scale, there is too little draft through the unit and an insufficient amount of combustion air is being drawn into the combustion zone. There may be a downdraft condition causing the products to be pushed back into the combustion zone, or there may be an obstruction in the venting system. The cause must be found and corrected.

With forced draft or induced draft equipment, a blocked vent usually results in a shutdown of the equipment. Also, the combustion blower

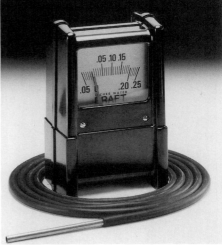

Figure 3-5. Two types of draft gauges (Courtesy, Bacharach, Inc.)

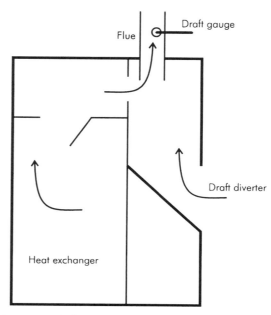

Figure 3-6. Checking vent draft

may be set too high, which causes excessive combustion air to be delivered to the unit. The reasons for both of these conditions must be found and corrected.

## Carbon Monoxide Analyzer

Carbon monoxide (CO) is a deadly, odorless gas that is produced during the combustion process. A carbon monoxide analyzer, sometimes called a monoxor, is used to measure the amount of CO present. The flue gas sample for this test must be taken at the inlet side of the draft diverter to make certain the sample is not diluted by air that is drawn into the draft diverter. The CO reading must never go above 0.04% in an air-free sample of the flue gas. CO in gas heating equipment flue gas is caused by improper combustion. The most common cause is either over-firing of the burners or insufficient primary air. These conditions are usually indicated by a yellow flame. The solution to these problems is to decrease the firing rate and increase the primary air to the main burners.

Insufficient secondary air will also cause a high CO reading, as well as a high carbon dioxide ($CO_2$) reading. If the flame shows no problems, check the $CO_2$ content and correct any problems found with the secondary air.

## Carbon Dioxide Analyzer

The total amount of secondary air for combustion purposes is designed into the unit during the design and manufacturing stages. The amount of secondary air is controlled by the use of baffles in the heat exchanger and/or flue outlet restrictions. The percentage of $CO_2$ and the temperature of the flue gases are indications of the percentage of combustion efficiency. Figure 3-7 shows a standard mechanical instrument for measuring $CO_2$.

Figure 3-7. Standard $CO_2$ analyzer (Courtesy, Bacharach, Inc.)

The $CO_2$ test sample should be taken on the inlet side of the draft diverter. The $CO_2$ in the flue gases indicates the amount of excess air passing through the combustion zone. It can also be considered as the amount of heat lost through the venting system.

The percentage of $CO_2$ in the flue gas products should be between 8.25% and 9.5%. Use the instrument manufacturer's guide to properly determine the percentage. If the $CO_2$ is not within the percentage range, an adjustment to the secondary air is required. As the amount of secondary air decreases, the amount of $CO_2$ decreases. Likewise, an increase in secondary air causes an increase in $CO_2$. Also, as the percentage of $CO_2$ decreases, the flue gas temperature increases, the wasted heat goes up and out through the vent system, and the combustion efficiency decreases.

## VELOMETER AND OTHER AIRFLOW MEASURING INSTRUMENTS

In air conditioning and refrigeration work, it is necessary that technicians understand how to determine airflow and what the readings indicate. Air velocity is the distance air travels in a given period of time. It is usually expressed in feet per minute (fpm). When the air velocity is multiplied by the cross section area of the duct, the volume of air flowing past that point in the duct can be determined. This volume of airflow is usually expressed in cubic feet per minute (cfm). The velocity or air volume measurements can be used to determine if the airflow system is operating properly, or if some repairs are needed.

Some airflow instruments, such as the air velocity meter in Figure 3-8, have a direct read-out in cfm, while others require some calculations to determine the cfm flow through the system. This is one factor that must be taken into account when purchasing the instrument to be used. The instrument manufacturer will usually include instructions on the proper use of the instrument. Be sure to follow these instructions to obtain proper readings.

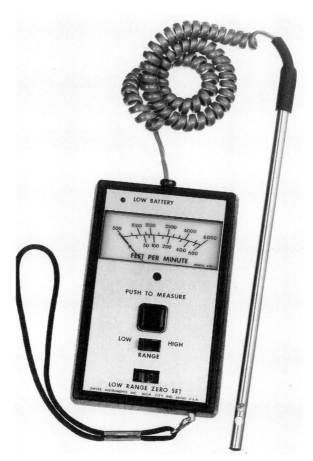

Figure 3-8. Air velocity meter (Courtesy, Dwyer Instruments, Inc.)

In use, the airflow instrument is held against the air outlet grille, and a reading is indicated on the meter scale, Figure 3-9. The reading is then interpreted according to the instrument manufacturer's instructions. When large grilles are used, the average of several readings is used to obtain the correct airflow. Use the instrument manufacturer's instructions on how to take these readings.

Figure 3-9. Proper use of an air velocity meter (Courtesy, Dwyer Instruments, Inc.)

## DETERMINING COMBUSTION EFFICIENCY

Use the following steps and the worksheet to determine the combustion efficiency of the heating unit:

1. Visually check the entire unit for cleanliness, and ensure all components are in proper working condition. Be sure the heat exchanger passages and the venting system are clear of all obstructions.
2. Determine the type of gas used.
3. Determine the Btu content of the gas per cubic foot. (Contact the local gas company or the LP gas delivery company.)
4. Start the heating unit, and allow it to operate for about ten minutes to bring everything up to operating temperature.
5. Measure the manifold gas pressure.

6. Determine the type of flame (yellow, yellow tip, blue, etc.).
7. Adjust the burner to cause the flame to burn blue.
8. Check the temperature rise of the circulating air through the unit.
9. Determine the blower cfm.
10. Check the flue gas temperature.
11. Check the $CO_2$ content of the flue gases.
12. Determine the operating efficiency. Use the instrument manufacturer's procedure.
13. Measure the vent draft.
14. Measure the CO content of the flue gases.
15. Set the fan control to start the fan at 130°F and stop the fan at 100°F.
16. Determine the unit cfm. Use the following formula:

$$cfm = \frac{Btu \times Combustion\ efficiency}{1.08 \times \Delta T}$$

When the heating unit is operating at its peak capacity but is not heating the structure properly, there are some other options available. First, open all the supply air grilles, measure the cfm from each one, and total them. If the cfm is 10% less than that determined in Steps 8 and 9, air is leaking from the duct system. Before the unit will operate efficiently the leak must be repaired.

The next step is to measure the duct heat loss. To do this, measure the discharge air temperature as it leaves the supply air grille located farthest from the heating unit. The difference in the discharge air temperature at the supply air grille and the temperature of the discharge air from the heating unit should not exceed 3° to 5°F. If the difference is more than 5°F, the duct needs more insulation.

When the heating system meets this criteria, the only other alternative is to install a larger heating unit. Do not increase the size of the main burner orifices or increase the firing rate of the furnace above the recommended input rating of the furnace manufacturer. To do so is very dangerous.

OPERATING AT PEAK EFFICIENCY

# Gas Heating Worksheet

Introduction: Use the following procedures and the test instrument manufacturers' instructions to determine the combustion efficiency of gas burning equipment.

Tools Needed: tool kit, gas manifold pressure gauge, dry bulb thermometer, flue gas temperature thermometer, draft gauge, carbon monoxide analyzer, carbon dioxide analyzer, and velometer.

Procedures:

1. Visually check the entire system for cleanliness, and ensure all components are in proper working condition. Be sure the heat exchanger passages and the venting system are clear of all obstructions.

2. Determine the type of gas (natural, LP) and record. _____

3. Determine the Btu content of the gas and record. _____ per cubic foot

4. Start the heating unit, and allow it to operate for about ten minutes.

5. Measure the manifold gas pressure and record. _____ inches w.c.

6. Determine the type of flame and record. _____

7. Adjust the burner if needed, and record the type of flame. _____

8. Determine the temperature rise of the circulating air through the unit and record. Use the following formula:

$$\Delta T = \text{Discharge air temperature - Entering air temperature}$$

$$\Delta T = \underline{\hspace{2cm}} °F$$

9. Determine the blower cfm and record. Use the following formula:

$$cfm = \frac{Btu}{1.08 \times \Delta T}$$

$$cfm = \underline{\hspace{2cm}}$$

10. Check the flue gas temperature and record. _____ °F

11. Check the $CO_2$ content of the flue gases and record. _____ %

12. Determine the operating combustion efficiency and record. _____ %

13. Measure the unit vent draft and record. _____ inches w.c.

14. Measure the CO content of the flue gases and record. _____ %

Copyright © Business News Publishing Company

15. Make any adjustments or repairs required to increase the combustion efficiency of the unit. Repeat Steps 3 through 13.

16. Is this what the manufacturer rates the unit? _____

> Multiple tear-out copies
> of this worksheet
> can be found in
> Appendix B.

# Chapter 4

# Oil Burners

When adjusting an oil burner, the main objective is to ensure the unit produces as much heat as possible with as little heat input as possible. The efficiency adjustment must provide the greatest heat gain without causing undesirable conditions. In mathematical terms, efficiency may be explained as the following:

$$\text{Efficiency} = \frac{\text{Useful energy out}}{\text{Total energy}}$$

The components must be clean and in good working condition, and clean air filters must be installed.

## Adjustment Process

There are several closely related factors involved in the adjustment process. When an adjustment is made to one of these factors, the other factors are also affected.

## Required Test Instruments

The proper test instruments are needed to make accurate measurements of the product or process being tested. The old procedure of merely looking at the equipment or feeling a line no longer indicates what is happening with the unit.

There are many brands and models of test instruments available. Which particular instrument to use is the user's choice, but accurate test instruments must be used to properly fine tune oil burners. The exact procedure for instrument use can be found in the manufacturer's operating instructions. These instruments must be properly cared for and their accuracy maintained for optimum performance.

This chapter will discuss the specific procedures used to efficiency test oil-fired heating equipment, not specific instruments. The following is a list of the basic test instruments required for combustion testing and adjusting the efficiency of an oil-fired unit:

1. Flue gas temperature thermometer
2. Draft gauge
3. Smoke tester
4. Carbon dioxide analyzer with the appropriate combustion efficiency chart or slide rule to use in combination with the various test results to determine the combustion efficiency
5. Dry bulb thermometer

### Flue Gas Temperature Thermometer

This type of thermometer is used for measuring higher than normal temperatures from about 200° to 1000°F (see Figure 3-4). They are used to measure the temperature of flue gases in the vent system, or the stack temperature. The flue gas temperature in oil-fired heating units should not be more than 630°F higher than the ambient or combustion air temperature. The minimum flue gas temperature should be 380°F.

High flue gas temperatures result in excessive heat loss up the chimney and excessive oil consumption and may be caused by the following:

- Dirty heat exchanger (possibly high smoke and high draft loss)
- Furnace over-firing and a high $CO_2$ content of the flue gases
- Poor furnace design (more baffles are needed to reduce combustion flow)
- Poor combustion chamber design
- Excessive draft through the unit combustion zone

Low flue gas temperatures may result in moisture condensation, rusting and deterioration of the smoke pipe and chimney, and poor draft. Low flue gas temperatures may also be caused by under-firing of the unit.

The flue gas temperature is taken by drilling a 1/4-in. hole in the flue pipe about 12 ft from the unit or boiler breaching on the unit side of the draft regulator and at least 6 in. away from the regulator, Figure 4-1.

Start the oil burner, and allow it to operate for about 15 minutes before taking a reading. Determine the net flue gas temperature by subtracting the ambient air temperature from the flue gas thermometer reading.

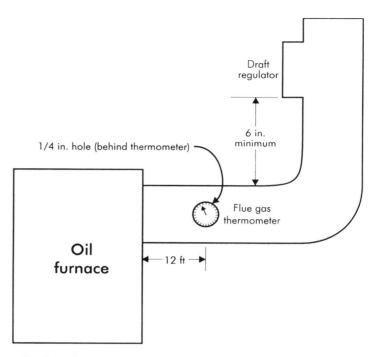

Figure 4-1. Checking flue gas temperature on an oil burner

## *Draft Gauge*

The correct amount of draft is essential for efficient oil burner operation. The draft itself is not directly related to combustion efficiency, but it does offset its efficiency. The amount of draft determines how fast the products of combustion pass through the combustion zone. The amount of draft also determines the amount of combustion air supplied to the burner. An excessive amount of draft can reduce the percentage of $CO_2$ in the flue gases and increase the temperature of the flue gases (see Figure 3-5 for two examples of draft gauges).

Each type of installation has its own draft requirements. Not enough draft can cause pressure in the combustion area, thus allowing smoke and odor to escape from the unit. A lack of draft can also make it impossible to adjust the oil burner to maximum efficiency, because maximum efficiency is dependent upon the proper mixture of air and oil during burner operation.

There are two kinds of draft to be checked and adjusted on oil burners: over-fire draft and flue pipe draft.

### Over-Fire Draft

An over-fire draft of at least 0.02 inches w.c. is considered sufficient to develop and maintain proper combustion. Should the over-fire draft fall below 0.02 inches w.c., smoke and oil odor may be present in the burner area. Also, a rumbling or pulsating condition may be present when close adjustments to obtain the greatest efficiency are completed.

### Flue Pipe Draft

The flue pipe drafts must be adjusted to prevent positive pressure in the combustion chamber. A long and complex flue passage requires a greater amount of flue pipe draft. A short and simple flue passage requires a lower flue pipe draft. Oil burners rated at 1.5 gph or less require a flue pipe draft of between 0.04 and 0.06 inches w.c. to maintain an over-fire draft of 0.02 inches w.c. in the combustion zone.

The flue pipe draft is adjusted by changing the position of the draft regulator damper. Adjusting the counterweight so the damper moves toward the closed position causes an increase in the draft. Adjusting the counterweight so the damper moves toward the open position causes a decrease in the draft through the combustion zone.

### *Smoke Tester*

The smoke tester and the smoke scale are used to determine the amount of smoke in the flue gases, Figure 4-2. Combustion gases containing an excessive amount of smoke indicate incomplete combustion and inefficient oil burner operation. An excessive amount of smoke allows soot formation on the heat exchanger surfaces. This soot slows down the heat transfer and clogs the flue passages, thus preventing proper draft through the combustion zone. A soot buildup of 1/8" can reduce heat transfer by up to 10%.

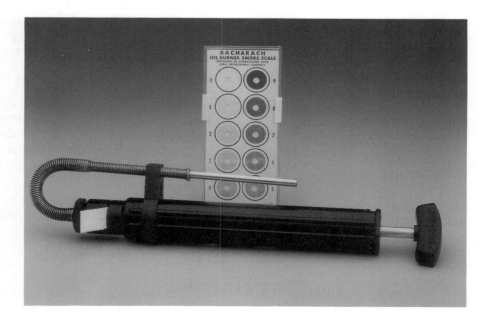

Figure 4-2. Smoke test kit (Courtesy, Bacharach, Inc.)

The purpose of conducting the smoke test is to determine the amount of smoke the flue gases contain. The smoke test can then be used along with other tests to adjust the oil burner to obtain maximum efficiency. The smoke scale has 10 color-graded spots ranging from 0 to 9, where 0 is pure white and 9 is the darkest color on the scale. The smoke scale is always used along with the smoke tester for complete results. Use the instrument manufacturer's instructions when conducting the smoke test.

Not all types of oil burners are affected the same way by the same amount of smoke in the flue gases. The type of combustion zone construction, to a great extent, determines how fast soot accumulates on the heat exchanger and other surfaces. Soot accumulates very rapidly on some types of construction when fired with a #3 smoke spot, whereas other units may accumulate soot at a much slower rate with the same smoke content. Table 4-1 shows the possible soot accumulation rate for different amounts of smoke in the flue gases.

## *Carbon Dioxide Analyzer*

The correct percentage of $CO_2$ in the flue gases is very important to oil burner combustion efficiency testing. The desired $CO_2$ reading is between 9% and 10%. When the reading is below 9%, check for the following conditions:

- Air leakage
- Excess combustion air
- Worn, plugged, or incorrect nozzles
- High draft conditions
- Incorrect or defective combustion zone
- Poor atomization of the fuel oil
- Incorrect combustion air handling parts
- Incorrectly set oil pressure

| Smoke scale | Rating | Soot accumulation |
| --- | --- | --- |
| 1 | Excellent | Reduced, very light accumulation if at all |
| 2 | Good | Soot accumulation light, will not appreciably increase stack temperature |
| 3 | Fair | May be some soot accumulation, will seldom require cleaning more than annually |
| 4 | Poor | Borderline condition, some units may require more than annual cleaning |
| 5 | Very poor | Soot accumulation very heavy and rapid |

Table 4-1. Smoke effect on burner performance

- Excessive combustion zone draft or air leaks
- Erratic draft regulator

There are two factors that determine the amount of heat lost in the flue gases. They are the flue gas temperature and the percentage of $CO_2$ in the flue gases. It is the flue gas heat loss that determines the combustion efficiency of the oil burner.

The $CO_2$ analyzer is an instrument that is used to sample flue gases to determine the $CO_2$ content (see Figure 3-7). A low $CO_2$ reading indicates that not all of the fuel is burning completely, thus some adjustments are required. It is recommended that the $CO_2$ analyzer be used along with other testing devices. One such device is the slide rule calculator, which can be used to determine the unit efficiency and stack losses in oil burner installations by correlating the temperature of the flue gases and the $CO_2$ percentage. To properly use the $CO_2$ analyzer, follow the instrument manufacturer's instructions.

When the $CO_2$ reading is above 10%, at least one of the following conditions must be corrected:

- Insufficient draft
- Oil pump not functioning properly
- Poor fuel supply
- Improper fuel/air ratio
- Draft regulator improperly adjusted
- Improper combustion fan delivery
- Defective or incorrect nozzle type
- Wrong burner air handling parts
- Excessive air leaks in the combustion zone

### Dry Bulb Thermometer

A dry bulb (db) thermometer is used for measuring air temperatures (see Figure 2-1). Determining the temperature rise of the air as it passes through the heating unit is a very important step in the fine tuning process. The dry bulb temperature is a measure of the amount of heat absorbed from the heat exchanger by the air. The air temperature is measured in two locations: the return air stream and the discharge air stream, Figure 4-3. Be sure to take the temperature readings after the mixing of any air. The thermometer must be placed where the radiant heat from the heat exchanger cannot be measured by the thermometer. Radiant heat can cause a faulty temperature reading and an incorrect test.

Two different thermometers that have been tested and found to produce exactly the same readings under the same set of circumstances should be used. If two thermometers that read exactly the same temperature are not available, then use one thermometer to measure both temperatures. An electronic thermometer that has separate, properly adjusted leads also produces the desired results. The best operating air temperature rise of an oil-fired unit is between 60° and 80°F. This allows the blower to run longer, distributing the heat more evenly through the building.

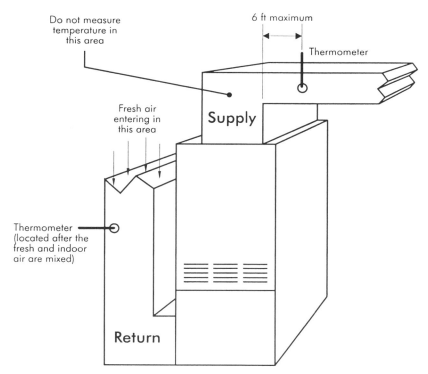

Figure 4-3. Measuring temperature rise through a furnace

## COMBUSTION EFFICIENCY RESULTS

Oil burner combustion efficiency is considered to be very good when a conversion oil burner installation is operating with a flue gas temperature of between 600° and 700°F. When a packaged oil burner unit has a flue gas temperature between 400° and 500°F, the combustion efficiency is also considered to be very good. However, these flue gas temperatures must be reached along with other required test readings, including the following: smoke scale reading of a #1 or #2 spot, with a #3 spot being the maximum allowed; a $CO_2$ reading of between 9% and 10%; and an over-fire draft of between 0.04 and 0.06 inches w.c. When any of these requirements are not met, some corrective measures must be taken to obtain maximum combustion efficiency.

## DETERMINING COMBUSTION EFFICIENCY

Use the following steps and the worksheet to determine the combustion efficiency of an oil burner:

1. Visually check the entire unit for cleanliness, and ensure all components are in proper working condition. Be sure the flue gas passages and the venting system are clear of all obstructions.
2. Start the oil burner, and allow it to operate for at least 15 minutes.

3. Check the flue gas temperature. For conversion oil burner installations, flue gas temperature should be between 600° and 700°F. For packaged unit installations, flue gas temperature should be between 400° and 500°F.
4. Check the $CO_2$ content of the flue gases. Flue gases should be between 9% and 10% $CO_2$.
5. Determine the operating combustion efficiency. Use the instrument manufacturer's procedure.
6. Conduct a smoke test. Smoke spot reading should be either #1 or #2 smoke spot. A #3 smoke spot should be considered the maximum.
7. Check the over-fire draft. It should be between 0.04 and 0.06 inches w.c.
8. Determine the unit cfm. Use the following formula:

$$cfm = \frac{gph \times Btu \times Combustion\ efficiency}{1.08 \times \Delta T}$$

When the heating unit is operating at its peak efficiency but is not heating the structure properly, there are some other options available. First, open all the supply air grilles, measure the cfm from each one, and total them. If the cfm is 10% less than that determined in Step 8, air is leaking from the duct system. Find the leak and repair it.

The next step is to measure the duct heat loss. To do this, measure the discharge air temperature as it leaves the supply air grille located farthest from the heating unit. The difference in the discharge air temperature at the supply air grille and the temperature of the discharge air from the heating unit should not exceed 3° to 5°F. If the difference is more than 5°F, the duct needs more insulation.

When the heating system meets this criteria, the only other alternative is to either increase the firing rate of the oil burner or install a larger capacity burner that matches the furnace combustion requirements. Check with various oil burner manufacturers for the correct unit.

# Oil Heating Worksheet

Introduction: When a customer complains about not enough heat from an oil-fired unit, the service technician should, as a first step, determine if the oil burner is delivering the amount of oil it was designed to deliver. Use the following procedures and the test instrument manufacturers' instructions to determine the combustion efficiency of oil burning equipment.

Tools Needed: tool kit, flue gas temperature thermometer, draft gauge, smoke tester, and carbon dioxide analyzer with the appropriate combustion efficiency chart or slide rule to use in combination with the various test results to determine the combustion efficiency.

Procedures:

1. Visually check the entire unit, and ensure all components are in proper working condition. Be sure the flue gas passages and the venting system are clear of all obstructions.

2. Is this a conversion burner? _____

3. Start the oil burner, and allow it to operate for about 15 minutes.

4. Check the flue gas temperature and record. _____°F

5. Check the $CO_2$ content of the flue gases and record. _____%

6. The operating combustion efficiency is _____%

7. Conduct the smoke test and record. _____spot

8. Measure the over-fire draft. _____inches w.c.

9. Make any adjustments or repairs required to increase the combustion efficiency of the unit. Repeat Steps 3 through 8.

10. Measure the unit cfm and record. _____cfm

11. Is this what the manufacturer rates the equipment? _____

> Multiple tear-out copies
> of this worksheet
> can be found in
> Appendix B.

Copyright © Business News Publishing Company

# Chapter 5

# Air Conditioning Systems and Heat Pumps (Cooling Mode)

Air conditioning systems are the largest users of electricity in both the home and most smaller commercial buildings. The cooling process involves sensible and latent heat removal. The air is cooled to the dew point temperature, removing only sensible heat. Both sensible and latent heat are removed below the dew point temperature. The removal of latent heat from the air requires a large part of the air conditioning unit's capacity. Therefore, fine tuning these systems results in lowering the consumption of electricity by as much as 25% in some cases. The percent by which the electrical consumption is lowered depends on the condition of the unit before it is fine tuned and the condition after the process is completed. There are several steps involved in the fine tuning process, and each step requires that all of the air conditioning system components be clean and in good working condition and that clean air filters have been installed.

## ADJUSTMENT PROCESS

There are several closely related factors involved in the adjustment process. When an adjustment is made to one of these factors, the other factors are also affected.

## REQUIRED TEST INSTRUMENTS

The proper test instruments are needed to make accurate measurements of the product or process being tested. The old procedure of merely looking at something or feeling a line to determine the operating characteristics no longer indicates what is happening with the unit.

There are many brands and models of test instruments available. Which particular instrument to use is the user's choice, but accurate test instruments must be used to properly fine tune air conditioning units. The exact procedure for instrument use can be found in the manufacturer's operating instructions. These instruments must be properly cared for and their accuracy maintained for optimum performance.

This chapter will discuss the specific procedures used to fine tune air conditioning units, not specific instruments. The following is a list of the basic test instruments required for testing and adjusting the efficiency of an air conditioning system:

1. Wet bulb thermometer
2. Dry bulb thermometer
3. Total heat chart (or a psychrometric chart if preferred)
4. Gauge manifold
5. Velometer, anemometer, or flowhood (as desired or available)

### Wet Bulb Thermometer

A wet bulb (wb) thermometer is the same as an ordinary dry bulb thermometer, except that the bulb is covered with a wet cloth or gauze. Wet bulb thermometers are used to indicate the total heat content of a quantity of air. The same thermometer is used for all readings. Wet bulb thermometers are usually incorporated into sling psychrometers, Figure 5-1. The wet bulb thermometer is exposed to a rapidly moving air stream, and the wet bulb temperature is taken after the same temperature is indicated for three consecutive readings. The temperature indicated by a wet bulb thermometer will be lower than that indicated by an ordinary dry bulb thermometer when used to test identical conditions.

The difference between the wet bulb and dry bulb readings is known as the *wet bulb depression*. Calculating the wet bulb depression allows the relative humidity to be determined. Using Table 5-1, when the dry bulb temperature in a room measures 60°F and the wet bulb temperature is 50°F, the depression is 10°F. The chart shows that the relative humidity is 49%. If the wet bulb depression is 0°F, the relative humidity is 100%, meaning the air is saturated with moisture.

### Dry Bulb Thermometer

A dry bulb (db) thermometer is an ordinary thermometer used to measure air temperatures (see Figure 2-1). Determining the temperature fall of air as it passes through the air conditioning system or heat pump is very important when fine tuning cooling systems. The dry bulb temperature is a measure of the amount of heat absorbed from the air by the cooling coil. The air temperature is measured in two locations: the return air stream and the discharge air stream, Figure 5-2.

AIR CONDITIONING SYSTEMS AND HEAT PUMPS

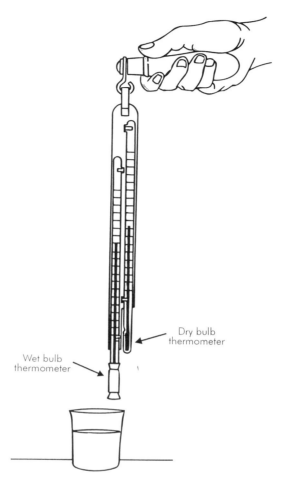

Figure 5-1. Sling psychrometer

| db temp. | wb depression |||||||||||||||||||||||||||||||
|---|---|---|---|---|---|---|---|---|---|---|---|---|---|---|---|---|---|---|---|---|---|---|---|---|---|---|---|---|---|---|
|  | 1 | 2 | 3 | 4 | 5 | 6 | 7 | 8 | 9 | 10 | 11 | 12 | 13 | 14 | 15 | 16 | 17 | 18 | 19 | 20 | 21 | 22 | 23 | 24 | 25 | 26 | 27 | 28 | 29 | 30 |
| 32 | 90 | 79 | 69 | 60 | 50 | 41 | 31 | 22 | 13 | 4 |  |  |  |  |  |  |  |  |  |  |  |  |  |  |  |  |  |  |  |  |
| 36 | 91 | 82 | 73 | 65 | 56 | 48 | 39 | 31 | 23 | 14 | 6 |  |  |  |  |  |  |  |  |  |  |  |  |  |  |  |  |  |  |  |
| 40 | 92 | 84 | 76 | 68 | 61 | 53 | 46 | 38 | 31 | 23 | 16 | 9 | 2 |  |  |  |  |  |  |  |  |  |  |  |  |  |  |  |  |  |
| 44 | 93 | 85 | 78 | 71 | 64 | 57 | 51 | 44 | 37 | 31 | 24 | 18 | 12 | 5 |  |  |  |  |  |  |  |  |  |  |  |  |  |  |  |  |
| 48 | 93 | 87 | 80 | 73 | 67 | 60 | 54 | 48 | 42 | 36 | 34 | 25 | 19 | 14 | 8 |  |  |  |  |  |  |  |  |  |  |  |  |  |  |  |
| 52 | 94 | 88 | 81 | 75 | 69 | 63 | 58 | 52 | 46 | 41 | 36 | 30 | 25 | 20 | 15 | 10 | 6 | 0 |  |  |  |  |  |  |  |  |  |  |  |  |
| 56 | 94 | 88 | 82 | 77 | 71 | 66 | 61 | 55 | 50 | 45 | 40 | 35 | 34 | 26 | 24 | 17 | 12 | 8 | 4 |  |  |  |  |  |  |  |  |  |  |  |
| 60 | 94 | 89 | 84 | 78 | 73 | 68 | 63 | 58 | 53 | 49 | 44 | 40 | 35 | 31 | 27 | 22 | 18 | 14 | 6 | 2 |  |  |  |  |  |  |  |  |  |  |
| 64 | 95 | 90 | 85 | 79 | 75 | 70 | 66 | 61 | 56 | 52 | 48 | 43 | 39 | 35 | 34 | 27 | 23 | 20 | 16 | 12 | 9 |  |  |  |  |  |  |  |  |  |
| 68 | 95 | 90 | 85 | 81 | 76 | 72 | 67 | 63 | 59 | 55 | 51 | 47 | 43 | 39 | 35 | 31 | 28 | 24 | 21 | 17 | 14 |  |  |  |  |  |  |  |  |  |
| 72 | 95 | 91 | 86 | 82 | 78 | 73 | 69 | 65 | 61 | 57 | 53 | 49 | 46 | 42 | 39 | 35 | 32 | 28 | 25 | 22 | 19 |  |  |  |  |  |  |  |  |  |
| 76 | 96 | 91 | 87 | 83 | 78 | 74 | 70 | 67 | 63 | 59 | 55 | 52 | 48 | 45 | 42 | 38 | 35 | 32 | 29 | 26 | 23 |  |  |  |  |  |  |  |  |  |
| 80 | 96 | 91 | 87 | 83 | 79 | 76 | 72 | 68 | 64 | 61 | 57 | 54 | 54 | 47 | 44 | 41 | 38 | 35 | 32 | 29 | 27 | 24 | 21 | 18 | 16 | 13 | 11 | 8 | 6 | 1 |
| 84 | 96 | 92 | 88 | 84 | 80 | 77 | 73 | 70 | 66 | 63 | 59 | 56 | 53 | 50 | 47 | 44 | 41 | 38 | 35 | 32 | 30 | 27 | 25 | 22 | 20 | 17 | 15 | 12 | 10 | 8 |
| 88 | 96 | 92 | 88 | 85 | 81 | 78 | 74 | 71 | 57 | 64 | 61 | 58 | 55 | 52 | 49 | 46 | 43 | 41 | 38 | 35 | 33 | 30 | 28 | 25 | 23 | 21 | 18 | 16 | 14 | 12 |
| 92 | 96 | 92 | 89 | 85 | 82 | 78 | 75 | 72 | 69 | 65 | 62 | 59 | 57 | 54 | 51 | 48 | 45 | 43 | 40 | 38 | 35 | 33 | 30 | 28 | 26 | 24 | 22 | 19 | 17 | 15 |
| 96 | 96 | 93 | 89 | 86 | 82 | 79 | 76 | 73 | 70 | 67 | 74 | 61 | 58 | 55 | 53 | 50 | 47 | 45 | 42 | 40 | 37 | 35 | 33 | 31 | 29 | 26 | 24 | 22 | 20 | 18 |
| 100 | 96 | 93 | 90 | 86 | 83 | 80 | 77 | 74 | 71 | 68 | 65 | 62 | 59 | 57 | 54 | 52 | 49 | 47 | 44 | 42 | 40 | 37 | 35 | 33 | 31 | 29 | 27 | 25 | 23 | 21 |
| 104 | 97 | 93 | 90 | 87 | 84 | 80 | 77 | 74 | 72 | 69 | 66 | 63 | 61 | 58 | 56 | 53 | 51 | 48 | 46 | 44 | 41 | 39 | 37 | 35 | 33 | 31 | 29 | 27 | 25 | 24 |
| 108 | 97 | 93 | 90 | 87 | 84 | 81 | 78 | 75 | 72 | 70 | 67 | 64 | 62 | 59 | 57 | 54 | 52 | 50 | 47 | 45 | 43 | 41 | 39 | 37 | 35 | 33 | 31 | 29 | 28 | 26 |

Table 5-1. Wet bulb depression chart

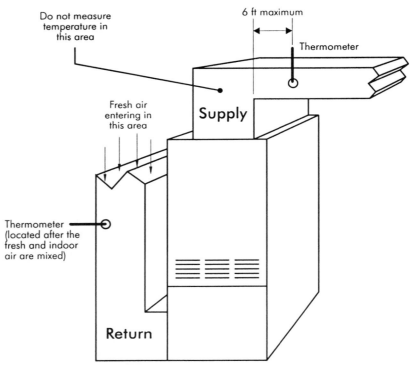

Figure 5-2. Measuring temperature drop across cooling coil

Two different thermometers that have been tested and found to produce exactly the same readings under the same set of circumstances should be used. If two thermometers that read exactly the same temperature are not available, then use one thermometer to measure both temperatures. It is best to measure the return air temperature first, because it will not change as fast as the discharge air temperature. An electronic thermometer with separate, properly adjusted leads also produces the desired results. Dry bulb thermometers are also incorporated into sling psychrometers.

## *Total Heat Chart*

A total heat ($H_t$) chart, also known as an enthalpy chart, is used to determine the total amount of heat in a quantity of air. The total heat chart in Table 5-2 lists the wet bulb temperatures (°F) in whole numbers down the left-hand column. The degrees are broken down into one-tenth increments across the top of the table. The amount of total heat in the air is listed in the columns underneath and to the right of the temperatures. For example, if the wet bulb temperature reading is 50.5°F, what is the total heat in the air? Find 50°F in the left-hand column, and follow that line to the right until the 0.5 column is reached. The total heat content of the air at 50.5°F wb is 20.58 Btu/lb of dry air.

| Temp. (°F wb) | .0 | .1 | .2 | .3 | .4 | .5 | .6 | .7 | .8 | .9 |
|---|---|---|---|---|---|---|---|---|---|---|
| 50 | 20.31 | 20.36 | 20.42 | 20.47 | 20.53 | 20.58 | 20.64 | 20.70 | 20.75 | 20.81 |
| 51 | 20.83 | 20.86 | 20.92 | 20.98 | 21.04 | 21.09 | 21.21 | 21.26 | 21.32 | 21.38 |
| 52 | 21.43 | 21.49 | 21.55 | 21.60 | 21.66 | 21.72 | 21.78 | 21.83 | 21.89 | 21.95 |
| 53 | 22.00 | 22.06 | 22.12 | 22.19 | 22.24 | 22.30 | 22.36 | 22.43 | 22.49 | 22.55 |
| 54 | 22.62 | 22.68 | 22.74 | 22.80 | 22.86 | 22.92 | 23.98 | 23.04 | 23.11 | 23.16 |
| 55 | 23.22 | 23.28 | 23.34 | 23.40 | 23.46 | 23.52 | 23.58 | 23.64 | 23.71 | 23.77 |
| 56 | 23.80 | 23.90 | 23.96 | 24.03 | 24.09 | 24.15 | 24.21 | 24.28 | 24.34 | 24.40 |
| 57 | 24.46 | 24.53 | 24.59 | 24.66 | 24.72 | 24.79 | 24.85 | 24.92 | 24.99 | 25.05 |
| 58 | 25.11 | 25.18 | 25.25 | 25.32 | 25.38 | 25.45 | 25.51 | 25.58 | 25.65 | 25.71 |
| 59 | 25.78 | 25.85 | 25.91 | 25.99 | 26.06 | 26.12 | 26.19 | 26.26 | 25.33 | 26.39 |
| 60 | 26.46 | 26.53 | 26.60 | 26.67 | 26.74 | 26.81 | 26.88 | 26.94 | 27.01 | 27.08 |
| 61 | 27.15 | 27.21 | 27.28 | 27.35 | 27.42 | 27.48 | 27.55 | 27.62 | 27.69 | 27.76 |
| 62 | 27.84 | 27.92 | 28.00 | 28.07 | 28.14 | 28.21 | 28.29 | 28.36 | 28.43 | 28.50 |
| 63 | 28.57 | 28.65 | 28.72 | 28.79 | 28.86 | 28.94 | 29.01 | 29.08 | 29.16 | 29.23 |
| 64 | 29.30 | 29.38 | 29.45 | 29.53 | 29.60 | 29.68 | 29.76 | 29.93 | 29.91 | 29.98 |
| 65 | 30.01 | 30.13 | 30.21 | 30.29 | 30.37 | 30.45 | 30.52 | 30.60 | 30.68 | 30.78 |
| 66 | 30.85 | 30.92 | 31.00 | 31.07 | 31.15 | 31.23 | 31.31 | 31.39 | 31.47 | 31.54 |
| 67 | 31.62 | 31.70 | 31.77 | 31.85 | 31.93 | 32.01 | 32.09 | 32.17 | 32.25 | 32.33 |
| 68 | 32.42 | 32.51 | 32.59 | 32.67 | 32.76 | 32.84 | 32.92 | 33.01 | 33.09 | 33.17 |
| 69 | 33.26 | 33.34 | 33.42 | 33.50 | 33.59 | 33.67 | 33.75 | 33.84 | 33.92 | 34.01 |
| 70 | 34.09 | 34.17 | 34.26 | 34.34 | 34.43 | 34.51 | 34.60 | 34.69 | 34.77 | 34.88 |
| 71 | 34.96 | 35.04 | 35.13 | 35.22 | 35.31 | 35.40 | 35.48 | 35.57 | 35.66 | 35.74 |
| 72 | 35.83 | 35.92 | 36.01 | 36.10 | 36.19 | 36.27 | 36.37 | 36.46 | 36.55 | 36.65 |
| 73 | 36.74 | 36.83 | 36.92 | 37.02 | 37.11 | 37.21 | 37.30 | 37.39 | 37.48 | 37.57 |
| 74 | 37.66 | 37.76 | 37.85 | 37.94 | 38.04 | 38.14 | 38.23 | 38.33 | 38.43 | 38.52 |
| 75 | 38.61 | 38.71 | 38.80 | 38.90 | 38.99 | 39.09 | 39.18 | 39.28 | 39.37 | 39.47 |
| 76 | 39.57 | 39.67 | 39.77 | 39.87 | 39.97 | 40.07 | 40.17 | 40.27 | 40.38 | 40.48 |
| 77 | 40.58 | 40.68 | 40.78 | 40.88 | 40.98 | 41.08 | 41.18 | 41.28 | 41.38 | 41.48 |
| 78 | 41.58 | 41.69 | 41.79 | 41.89 | 42.00 | 42.10 | 42.20 | 42.31 | 42.41 | 42.52 |
| 79 | 42.63 | 42.73 | 42.83 | 42.94 | 43.05 | 43.15 | 43.26 | 43.37 | 43.48 | 43.59 |
| 80 | 43.70 | 43.81 | 43.91 | 44.02 | 44.13 | 44.24 | 44.36 | 44.46 | 44.57 | 44.68 |
| 81 | 44.78 | 44.89 | 45.00 | 45.11 | 45.23 | 45.34 | 45.45 | 45.57 | 45.68 | 45.80 |
| 82 | 45.91 | 46.02 | 46.13 | 46.24 | 46.35 | 46.47 | 46.58 | 46.69 | 46.71 | 46.82 |
| 83 | 46.99 | 47.16 | 47.28 | 47.40 | 47.52 | 47.63 | 47.75 | 47.87 | 47.99 | 48.10 |
| 84 | 48.22 | 48.34 | 48.46 | 48.58 | 48.70 | 48.82 | 48.94 | 49.06 | 49.19 | 49.31 |
| 85 | 49.43 | 49.56 | 49.68 | 49.80 | 49.92 | 50.05 | 50.17 | 50.29 | 50.41 | 50.54 |
| 86 | 50.66 | 50.79 | 50.91 | 51.04 | 51.16 | 51.29 | 51.42 | 51.55 | 51.67 | 51.80 |
| 87 | 51.93 | 52.06 | 52.19 | 52.32 | 52.45 | 52.58 | 52.71 | 52.84 | 52.97 | 53.09 |
| 88 | 53.22 | 53.35 | 53.48 | 53.62 | 53.75 | 53.89 | 54.02 | 54.15 | 54.29 | 54.42 |
| 89 | 54.56 | 54.70 | 54.83 | 54.97 | 55.10 | 55.24 | 55.38 | 55.52 | 55.66 | 55.80 |
| 90 | 55.93 | 56.07 | 56.20 | 56.34 | 56.48 | 56.62 | 56.76 | 56.91 | 57.05 | 57.19 |

Table 5-2. Total heat content of air (Btu per pound of dry air with vapor to saturate it)

## Gauge Manifold

The gauge manifold is used to perform many procedures involving refrigerant, oil, and evacuation of the refrigeration system. Gauge manifolds are made up of a pressure gauge, compound gauge, valve manifold, and hoses and connections, Figure 5-3. The operating refrigerant pressures can be directly related to temperature by the use of a pressure-temperature chart. Most gauges incorporate a corresponding temperature for the different pressures of various refrigerants. The manifold has ports for both the compound and pressure gauges. There

# Operating at Peak Efficiency

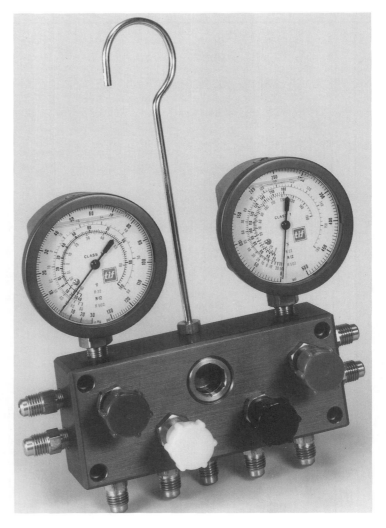

Figure 5-3. Gauge manifold (Courtesy, TIF Instruments, Inc.)

are valves in the manifold that are operated by hand and used to control the flow of fluids through the gauge ports and the charging hose ports.

## *Velometer, Anemometer, or Flowhood*

These very sensitive instruments are used to measure the velocity of air past a given point. Figures 5-4, 5-5, and 5-6 show examples of these various airflow measuring instruments. The air velocity readings measured by these instruments can be converted into cfm by using the following formula:

$$cfm = Area \times Velocity$$

The area must be calculated according to the instrument manufacturer's instructions or by multiplying the length of the coil by the width and then checking the equipment manufacturer's specifications for the free area of the coil. The number of readings taken over the area is usually recommended by the instrument manufacturer. These reading are then averaged. The average is found by using the following formula:

$$\text{Average} = \frac{\text{Total of readings}}{\text{Number of readings}}$$

Figure 5-4. Velometer (Courtesy, Alnor Instrument Company)

# Air-to-Air Heat Pump Units

At present, air-to-air heat pump units are the most popular types of heat pump systems. These units can provide both heating and cooling to the structure. Air-to-air heat pump units use air as the heat source during the heating phase and air for cooling the outdoor coil during the cooling phase.

## *Indoor Unit*

The indoor unit, which is comprised of the evaporator and blower, is the component that delivers the cool air to the space being conditioned.

## Operating at Peak Efficiency

Figure 5-5. Two different types of anemometers (Courtesy, Airflow Technical Products, Inc.)

Figure 5-6. Flowhood (Courtesy, Airflow Technical Products, Inc.)

Therefore, certain criteria must be met before the unit can operate as it was designed. Ensure that the following components are functioning properly:

- Refrigerant flow control
- Indoor blower
- Air delivery system (cleanliness)

### Refrigerant Flow Control

The refrigerant flow control is usually either a capillary tube or thermostatic expansion valve (TXV). In most cases, there are no adjustments possible to the capillary tube. The only adjustment for these devices is usually in the refrigerant charge; however, if the capillary tube is plugged or partially plugged, it should be replaced.

When a TXV is used, the superheat setting must be checked to ensure the correct amount for the specific application. Since air conditioning is considered to be a high temperature application, the superheat should be set at about 8° to 10°F. This will sufficiently feed the coil with liquid refrigerant and provide maximum operation.

### Indoor Blower

The indoor blower and the air delivery part of the system also contribute to the system operating at peak efficiency. The indoor air delivery system is designed to deliver about 400 cfm per ton of cooling. This can be measured with a velometer or flowhood. Also, the air flowing through the indoor coil should have a temperature drop of about 18° to 22°F db. Most units are designed to have a 20°F drop across the indoor coil.

### Air Delivery System (Cleanliness)

The filter, blower, indoor coil, and ductwork must be clean enough to allow the proper amount of air to flow through the system. This means that each of these items must be checked and cleaned before any efficiency testing begins.

## Outdoor Unit

There are generally two major components located in the outdoor unit: the compressor and the condenser. Each of these components plays a major part in the operation of the cooling system. The heat taken from the air flowing through the indoor coil is delivered by the compressor to the condensing coil where it is dissipated to the ambient air.

### Compressor

In effect, the compressor is the component that causes the refrigerant to circulate through the system. If the compressor is not functioning properly, the remainder of the system will not function as designed.

The compressor should be able to pump a vacuum on its low side of about 20 inches Hg. It should be able to hold this vacuum for at least five minutes against a normal discharge pressure. If the compressor does not meet this criteria, it will need to be replaced with one that does before full efficiency can be realized.

During normal operation, older units should have an operating suction pressure of about 70 psig when the indoor return air temperature is 80°F db. The high efficiency type units usually have a suction pressure of about 90 psig at this return air temperature. When servicing high efficiency units, it is best to use the unit manufacturer's data sheet for that particular make and model in order to determine the operating pressures. The proper pressure and its corresponding ambient temperature will be listed on the manufacturer's data sheet. Most operating pressures are different from one unit make to another, and some even vary from model to model within the same make.

### Condenser/Outdoor Coil

The condenser is where the refrigerant is cooled to the condensing point. If there is a factor present to prevent this condensation, the unit will not operate satisfactorily. The condenser must be clean and free of all debris. Condensers are designed to operate with an air temperature rise of about 25° to 35°F. If the temperature rise is out of this range, the cause must be found and corrected. When measuring the temperature in the condenser, be sure not to touch the coil or any metal with the thermometer or take the temperature in direct sunlight. Temperatures taken in such areas will not be accurate because of the radiant heat from the sun. Most condensers are designed to have an airflow of about 1000 cfm per ton of cooling capacity.

The condenser also provides the much needed subcooling of the liquid refrigerant. On most older model units, the desired liquid subcooling is about 10°F. The high efficiency models sometimes require a different subcooling temperature. Be sure to check the unit manufacturer's specifications for the unit make and model being serviced for the most accurate information.

The operating refrigerant discharge pressure usually corresponds to a saturation temperature of about 30°F above the ambient temperature.

## GEOTHERMAL HEAT PUMP UNITS

Geothermal heat pump units are one version of the water-to-air cooling system, except they provide both heating <u>and</u> cooling. Both the condenser cooling for the cooling operation and the heat source for the heating operation are provided with a water-source coil.

### *Indoor Coil*

The indoor coil is an air-over type. During the cooling operation, there should be a temperature drop across the indoor coil of about 18° to 22°F db. During the heating cycle, there should be a temperature rise across the indoor coil of about 50°F db. Check the equipment manufacturer's specifications for the exact readings.

### *Outdoor Coil*

The outdoor coil is water-cooled during cooling operation. It also uses water as the heat source for heating operation. There is a water valve for each of the cycles. Each valve must be set independently of the other. For cooling operation, the valve should be set to maintain a compressor discharge pressure of about 210 psig for R-22, or a condensing temperature of about 105°F. During heating operation, the heating water valve should be set to maintain a suction pressure of about 70 psig, or a suction temperature of 40°F. Check the equipment manufacturer's specifications for the exact settings.

### *Compressor*

The compressor on these units should be able to meet the previously outlined standards.

## DETERMINING UNIT EFFICIENCY

Use the following steps and the worksheet to determine the cooling capacity of the cooling unit:

1. Visually check the entire unit for cleanliness, and ensure all components are in proper working condition. Be sure the outdoor coils, the indoor coils, the indoor air filter, and the blower are clean.
2. Set the thermostat to demand cooling. Allow the unit to operate for about 15 minutes to allow the pressures and temperatures to stabilize.
3. Take the wet bulb and dry bulb temperatures of the air to the indoor coil. Use the same thermometer for both readings or two thermometers that read exactly the same.
4. Take the wet bulb and dry bulb temperatures of the air from the indoor coil. Use the same thermometer for both readings or two thermometers that read exactly the same.
5. Determine the total heat (enthalpy) output of the indoor coil. Use Table 5-2 or a psychrometric chart.

6. Determine the total heat (enthalpy) of the outlet air.
7. Determine the total heat (enthalpy) difference. Use the following formula:

$$\Delta H = \text{Inlet } H_t - \text{Outlet } H_t$$

For example, if the wet bulb temperature of the inlet air is 60°F and the outlet wet bulb temperature is 55°F, what is the difference in enthalpy? Using the total heat (enthalpy) formula and Table 5-2:

$$\Delta H = 26.46 - 23.22 = 3.24 \text{ Btu/lb of dry air}$$

8. Measure the evaporator free area in square feet.
9. Measure the velocity of the air flowing through the indoor coil.
10. Determine the indoor coil cfm. Use the following formula:

$$\text{cfm} = \text{Area} \times \text{Velocity}$$

11. Determine the unit capacity. Use the following formula:

$$\text{Btuh} = \text{cfm} \times 4.5 \times \Delta H$$

12. If this is not very close to the manufacturer's specifications, the cause must be found and corrected.

When the air conditioning unit or heat pump is operating at its peak efficiency but is not cooling the structure properly, there are some other options available. First, open all the supply air grilles, measure the cfm from each one, and total them. If the cfm is 10% less than that determined in Step 10, air is leaking from the duct system. Find the leak and repair it.

The next step is to measure the duct heat gain. To do this, measure the supply air temperature as it leaves the supply air grille located farthest from the cooling unit. The difference in the supply air temperature at the supply air grille and the temperature of the supply air from the cooling unit should not exceed 3° to 5°F. If the difference is more than 5°F, the duct needs more insulation.

When the cooling system meets this criteria, the only other alternative is to increase the size of the cooling unit or the cooling capacity of the heat pump system. Make sure the ductwork will handle the airflow required for the larger heat pump system.

# Air Conditioning and Heat Pump (Cooling Mode) Worksheet

Introduction: Use the following procedures and the test instrument manufacturers' instructions to determine the capacity of a cooling system.

Tools Needed: wet bulb thermometer, dry bulb thermometer, total heat content of air (Table 5-2) or psychrometric chart, tool kit, and velometer.

Procedures:

1. Set the thermostat to demand cooling.

2. Allow the unit to operate for about 15 minutes to allow the pressures and temperatures to stabilize.

3. Take the following temperature readings:

    Indoor coil:

    - Inlet air temperature (db) _____ °F
    - Inlet air temperature (wb) _____ °F
    - Outlet air temperature (db) _____ °F
    - Outlet air temperature (wb) _____ °F

    Outdoor coil:

    - Inlet air temperature (db) _____ °F
    - Outlet air temperature (db) _____ °F

4. Subtract the outlet air temperature (db) from the inlet air temperature (db) on the indoor coil and record. _____ °F

5. Subtract the inlet air temperature (db) from the outlet air temperature (db) on the outdoor coil and record. _____ °F

6. Using a psychrometric chart or Table 5-2, determine the total heat content (enthalpy) of the inlet air and record. _____ Btu/lb of dry air

7. Using a psychrometric chart or Table 5-2, determine the total heat content (enthalpy) of the outlet air and record. _____ Btu/lb of dry air

8. Determine the total heat (enthalpy) difference and record. Use the following formula:

$$\Delta H = \text{Inlet } H_t - \text{Outlet } H_t$$

$$\Delta H = \underline{\qquad} \text{ Btu/lb of dry air}$$

Copyright © Business News Publishing Company

OPERATING AT PEAK EFFICIENCY

9. Measure the evaporator free area in square feet and record. _____ ft²

10. Measure the air velocity through the indoor coil and record. _____

11. Determine the indoor coil cfm and record. Use the following formula:

        cfm = Area x Velocity

  cfm = _____

12. Determine the unit capacity and record. Use the following formula:

        Btuh = cfm x 4.5 x ΔH

  Btuh = _____

13. Is this what the manufacturer rates the equipment? _____

> Multiple tear-out copies
> of this worksheet
> can be found in
> Appendix B.

Copyright © Business News Publishing Company

# Chapter 6

# Heat Pumps (Heating Mode)

Fine tuning heat pumps in the heating mode is not difficult. In the heating mode, only sensible heat is added to the air as it passes through the unit. Therefore, dry bulb temperature measurements indicate the total performance of the unit. However, these units still must be thoroughly checked, and any required adjustments to the indoor airflow or refrigerant charge must be made in order for the unit to operate as efficiently as possible.

## REQUIRED TEST INSTRUMENTS

The proper test instruments are needed to make accurate measurements of the product or process being tested. The old procedure of merely looking at the equipment or feeling a line to determine the operating characteristics no longer indicates what is happening with the unit.

There are many brands and models of test instruments available. Which particular instrument to use is the user's choice, but accurate test instruments must be used to properly fine tune heat pumps in the heating mode. The exact procedure for instrument use can be found in the manufacturer's operating instructions. These instruments must be properly cared for and their accuracy maintained for optimum performance.

This chapter will discuss only those procedures used for efficiency testing heat pumps in the heating mode, not specific instruments. The following is a list of the basic test instruments required for testing and adjusting the efficiency of a heat pump operating in the heating mode:

1. Dry bulb thermometer
2. Velometer, anemometer, or flowhood (as desired or available)
3. Gauge manifold

# Operating at Peak Efficiency

4. Ammeter
5. Voltmeter
6. Wattmeter

### *Dry Bulb Thermometer*

A dry bulb (db) thermometer is an ordinary thermometer used to measure air temperature (see Figure 2-1). Determining the temperature rise of air as it passes through the heat pump unit is a very important step in the fine tuning process. The dry bulb temperature is a measure of the heat given up by the indoor coil to the air. The air temperature is measured in two locations: the return air stream and the discharge air stream, Figure 6-1.

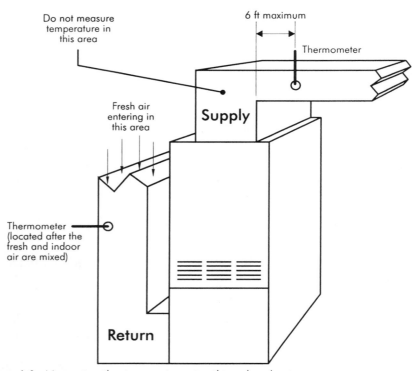

Figure 6-1. Measuring the temperature rise through a heat pump

Two different dry bulb thermometers that have been tested and found to produce exactly the same readings under the same set of circumstances can be used. If two thermometers that read exactly the same temperature are not available, then use one thermometer to measure both temperatures. It is best to measure the return air temperature first, because it will not change as fast as the discharge air. An electronic thermometer with two separate, properly adjusted leads also produces the desired results. Dry bulb thermometers are sometimes incorporated into sling psychrometers.

### Velometer, Anemometer, or Flowhood

These very sensitive instruments are used to measure the velocity of air past a given point (see Figures 5-4, 5-5, and 5-6). This reading can then be converted into cfm by using the following formula:

$$\text{cfm} = \text{Area} \times \text{Velocity}$$

The area must be calculated according to the instrument manufacturer's instructions or by multiplying the length of the coil by the width and then checking the equipment manufacturer's specifications for the free area of the coil. The number of readings taken over the area is usually recommended by the instrument manufacturer. These readings are then averaged. The average is found by using the following formula:

$$\text{Average} = \frac{\text{Total of readings}}{\text{Number of readings}}$$

### Gauge Manifold

The gauge manifold is used to perform many procedures involving refrigerant, oil, and evacuation of the refrigeration system. Gauge manifolds are made up of a pressure gauge, compound gauge, valve manifold, and the necessary hoses and connections (see Figure 5-3). The operating refrigerant pressures can be directly related to temperature by the use of a pressure-temperature chart. Most gauges incorporate a corresponding temperature for the different pressures of various refrigerants. The manifold has ports for both the compound and the pressure gauges. There are valves in the manifold that are operated by hand and used to control the flow of fluids through the gauge ports and the charging hose ports.

### Ammeter

The ammeter is used to determine the amount of electrical current used by the unit. The clamp-on type is the most popular, because the amperage can be measured without separating the wire (see Figure 2-3). These instruments are the most accurate when the wire being measured is in the center of the tongs.

### Voltmeter

The voltmeter is used to measure the voltage in a wire (see Figure 2-4). The analog-type meter is the most accurate when the indicator is in the center of the scale. The leads should be checked regularly to ensure they are in good working condition. When the insulation becomes worn or cracked, the leads should be replaced. A firm, solid fit between the meter and the lead must be maintained. A loose fitting lead can give an improper voltage reading.

### *Wattmeter*

The wattmeter is used to measure the voltage to the unit and the total wattage used by the unit (see Figure 2-5). These instruments are usually more accurate than using the voltmeter and the ammeter to determine the total wattage used by the unit. However, wattmeters are more expensive. The analog-type wattmeter is the most accurate when the indicator is reading at the midscale point.

## AIR-TO-AIR HEAT PUMP UNITS

Air-to-air heat pump units have air flowing over both the indoor coil and the outdoor coil. They have some unique operating characteristics, which are mostly determined by the outdoor ambient temperature. As with any other type of system, all components must be clean and in proper working condition. If not, any attempt to fine tune the system will be wasted.

### *Indoor Unit*

The indoor coil supplies heat to the structure during the heating cycle. These units are usually designed to operate with 400 to 500 cfm per ton of heating capacity. The airflow through these units is very critical. Airflow should be one of the first items checked when fine tuning a heat pump. The temperature of the air flowing through the indoor coil varies with the outdoor ambient temperature. However, the temperature is usually between 90° and 125°F. Check the equipment manufacturer's specifications for the unit being serviced for the exact discharge temperature desired for that unit.

### *Outdoor Unit*

The outdoor unit houses the compressor and outdoor coil, along with the other required components. The outdoor coil absorbs the heat from the ambient air so the refrigerant can take it indoors. The temperature of the outdoor ambient air determines both the suction and discharge refrigerant pressures. It is almost impossible to determine exactly what the proper pressures should be without the equipment manufacturer's specifications. However, the higher the outdoor ambient temperature, the higher the refrigerant pressures.

## GEOTHERMAL HEAT PUMP UNITS

Geothermal heat pump units are one version of the water-to-air cooling system, except they provide both heating <u>and</u> cooling. Both the condenser for the cooling operation and the heat source for the heating operation are provided with a water-source coil.

### *Indoor Coil*

The indoor coil is an air-over type. During the cooling operation, there should be a temperature drop across the indoor coil of about 18° to 22°F db. During the heating cycle, there should be a temperature rise across the coil of about 50°F db. Check the equipment manufacturer's specifications for the exact readings.

### *Outdoor Coil*

The outdoor coil is water-cooled during cooling operation. It also uses water as the heat source for heating operation. There is a water valve for each of the cycles. Each valve must be set independently of the other. For cooling operation, the valve should be set to maintain a compressor discharge pressure of about 210 psig for R-22, or a condensing temperature of about 105°F. During heating operation, the heating water valve should be set to maintain a suction pressure of about 70 psig, or a suction temperature of 40°F. Check the equipment manufacturer's specifications for the exact settings.

### *Compressor*

The compressor on these units should be able to pump a vacuum of at least 20 inches Hg when subjected to a normal discharge pressure. It should be able to hold this vacuum for about 5 minutes. If not, the valves are leaking and either the compressor or the valves must be replaced. It is very critical that these compressors meet this requirement because of the very severe conditions in which they operate. Some equipment manufacturers may require a different efficiency test for their compressor. Be sure to check their requirements.

## DETERMINING UNIT EFFICIENCY

Use the following steps and the worksheet to determine the heating capacity of a heat pump system:

1. Visually check the entire system for cleanliness, and ensure all components are in proper working condition. Be sure the indoor and outdoor coils, the indoor filter, and indoor blower are clean.
2. Set the thermostat to demand heat pump heating only. It may be necessary to disconnect the auxiliary heat strips so that only the heat pump can operate. Allow the unit to operate for about ten

minutes to allow the system pressures and temperatures to stabilize.
3. Check the refrigerant pressures on the operating system.
4. Check both the indoor and the outdoor dry bulb temperatures.
5. Compare the temperature and pressure readings to the equipment manufacturer's specifications.
6. Measure the entering and leaving air dry bulb temperatures across the indoor unit.
7. Determine the indoor unit cfm. Use either an airflow measuring meter as described above, or disable the heat pump unit and set the thermostat to bring on the auxiliary heat strips. If the auxiliary heat strip disconnect was turned off, be sure that it is turned back on. Then use the method described in Chapter 2 to determine the airflow. After determining the indoor cfm, turn off the auxiliary heat strips and turn on the heat pump unit.
8. Determine the heating capacity (Btu output) of the unit. Use the following formula:

$$\text{Heating capacity} = \text{cfm} \times 1.08 \times \Delta T$$

9. Determine the system coefficient of performance (COP) by measuring the watts used by the unit, including the indoor fan motor. Use the following formula:

$$COP = \frac{\text{Btu output}}{\text{Btu input}}$$

or:

$$COP = \frac{\text{Btu capacity}}{\text{Watts} \times 3.413}$$

For example, if a heat pump has a measured heating capacity of 40,000 Btuh and the measured wattage is 4.380 kW, what is the COP?

First, convert 4.380 kW to watts:

$$4.380 \times 1000 = 4380 \text{ W}$$

Then use the formula outlined earlier:

$$COP = \frac{40,000 \text{ Btu}}{4380 \text{ W} \times 3.413} = 2.67$$

When the heat pump unit is operating at its peak efficiency but is not heating the structure properly, there are some other options available. First, open all of the supply air grilles, measure the cfm from each one, and total them. If the cfm is 10% less than that determined in Step 8, air is leaking from the duct system. Find the leak and repair it.

The next step is to measure the duct heat loss. To do this, measure the supply air temperature as it leaves the supply air grille located farthest from the indoor unit. The difference in the supply air temperature at the supply air grille and the temperature of the supply air from the indoor unit should not exceed 3° to 5°F. If the difference is more than 5°F, the duct needs more insulation.

When the heat pump system meets this criteria, the only other alternative is to increase the size of the heat pump unit or increase the heating capacity of the auxiliary heat strips. Be aware that the electric bill will increase with an increase in heat strip capacity. Make sure the ductwork will handle the required airflow.

# Heat Pump (Heating Mode) Worksheet

Introduction: The service technician is often called upon to perform a capacity check of a heat pump system. This request may come in a variety of forms, such as a complaint of a high electric bill or not enough heat. Determining the capacity of a heat pump in the heating mode is not very difficult, and every service technician should perform this test whenever there is a suspected problem with the unit.

Tools Needed: dry bulb thermometer, velometer, gauge manifold, voltmeter, wattmeter, and tool kit.

Procedures:

1. Set the room thermostat to the heating or automatic position.

2. Set the temperature lever to demand heating.

3. Turn off all electricity to the auxiliary heat strips. Only the heat pump and the indoor fan are to be operating during this test.

4. Measure the temperature rise of the air through the indoor unit. Use the following formula:

$$\Delta T = \text{Leaving air temperature} - \text{Entering air temperature}$$

   a. Use the same thermometer for measuring both the return and supply air temperatures.

   b. Do not measure the temperature "in view" of the indoor coil. True air temperature cannot be measured in areas affected by radiant heat.

   c. Make the temperature measurements within 6 ft of the indoor unit. Measurements taken at the supply and return air grilles are not accurate enough.

   d. Use the average temperature when more than one duct is connected to the plenum. Use the following formula:

$$\text{Average} = \frac{\text{Total of readings}}{\text{Number of readings}}$$

   e. Make sure the air temperature is stable before taking these measurements.

   f. Take these measurements downstream from any mixed air source.

   g. Record the temperature difference of the return and supply air as $\Delta T$.

$$\Delta T = \underline{\qquad} - \underline{\qquad}$$

Copyright © Business News Publishing Company

5.  Determine the Btu output. Use the following formula:

$$Btu = cfm \times 1.08 \times \Delta T$$

where:
$$cfm = \frac{Btuh}{1.08 \times \Delta T}$$

1.08 = specific heat of air constant

$\Delta T$ = supply air temperature minus return air temperature

6.  Is this what the manufacturer rates the equipment? _____

> Multiple tear-out copies of this worksheet can be found in Appendix B.

# Chapter 7

# Refrigeration

Refrigeration is one of the major expenses of food stores, restaurants, processing plants, etc. Because of this expense, the owners and operators of these types of businesses are always interested in having the equipment operate as economically as possible. Technicians who are capable of fine tuning the refrigeration systems used in these applications will always be in demand and can charge more for their services. There are several steps involved in the fine tuning of refrigeration systems. All of these steps require that the system components be clean and in good working condition and that the system contain the proper refrigerant charge.

## ADJUSTMENT PROCESS

There are several closely related factors involved in the adjustment process. When an adjustment is made to one of these factors, the other factors are also affected.

## REQUIRED TEST INSTRUMENTS

The proper test instruments are needed to make accurate measurements of the product or process being tested. The old procedure of looking at equipment or feeling a line to determine the operating characteristics no longer indicates what is happening with the unit.

There are many brands and models of test instruments available. Which particular instrument to use is the user's choice, but accurate test instruments must be used to properly fine tune commercial refrigeration systems. The exact procedure for instrument use can be found in the manufacturer's operating instructions. These instruments must be properly cared for and their accuracy maintained for optimum performance.

This chapter will discuss only those procedures used for efficiency testing refrigeration systems, not specific instruments. The following is a list of the basic test instruments required for testing and adjusting the efficiency of a refrigeration system:

1. Dry bulb thermometer
2. Gauge manifold
3. Voltmeter
4. Ammeter
5. Velometer, anemometer, or flowhood (as desired or available)

### Dry Bulb Thermometer

A dry bulb (db) thermometer is an ordinary thermometer used to measure air temperature (see Figure 2-1). Determining the temperature drop of air as it passes through the evaporator or the temperature rise as it passes over the condenser is very important when fine tuning refrigeration systems. The dry bulb temperature is a measure of the heat given up to the air or taken in by the coil from the air. The air temperatures are measured in two locations: the return air stream and the discharge air stream, Figure 7-1.

Two different dry bulb thermometers that have been tested and found to produce exactly the same readings under the same set of circumstances should be used. If two thermometers that read exactly the same temperature are not available, then use one thermometer to measure both temperatures. It is best to measure the return air temperature first, because it will not change as fast as the discharge air temperature. An electronic thermometer with two separate, properly adjusted leads also produces the desired results.

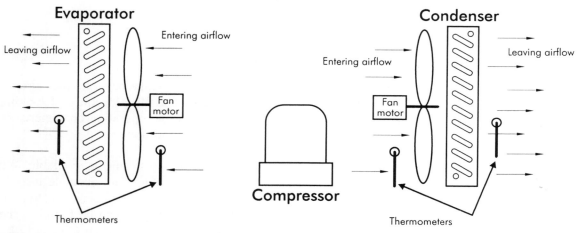

Figure 7-1. Measuring air temperature in a refrigeration system

### Gauge Manifold

The gauge manifold is used to perform many procedures involving refrigerant, oil, and evacuation of the refrigeration system. Gauge manifolds are made up of a pressure gauge, a compound gauge, a valve manifold, and the necessary hoses and connections (see Figure 5-3). The operating refrigerant pressures can be directly related to temperature by the use of a pressure-temperature chart. Most gauges incorporate a corresponding temperature for the different pressures of various refrigerants. The manifold has ports for both the compound and the pressure gauges. There are valves in the manifold that are operated by hand and used to control the flow of fluids through the gauge ports and the hose ports.

### Voltmeter

The voltmeter is used to measure the voltage in a wire (see Figure 2-4). The analog-type meter is the most accurate when the indicator is in the center of the scale. The leads should be checked regularly to ensure they are in good working condition. When the insulation becomes worn or cracked, the leads should be replaced. A firm, solid fit between the meter and the lead must be maintained. A loose fitting lead can give an improper voltage reading.

### Ammeter

The ammeter is used to determine the amount of electrical current used by the unit. The clamp-on type is the most popular, because the amperage can be measured without separating the wire (see Figure 2-3). These instruments are the most accurate when the wire being measured is in the center of the tongs.

### Velometer, Anemometer, or Flowhood

These very sensitive instruments are used to measure the velocity of air past a given point (see Figures 5-4, 5-5, and 5-6). This reading can then be converted into cfm by using the following formula:

$$cfm = Area \times Velocity$$

The area must be calculated according to the instrument manufacturer's instructions or by multiplying the length of the coil by the width and checking the equipment manufacturer's specifications for the free area of the coil. The number of readings taken over the area is usually recommended by the instrument manufacturer. These readings are then averaged. The average is found by using the following formula:

$$\text{Average} = \frac{\text{Total of readings}}{\text{Number of readings}}$$

## EVAPORATOR

Refrigeration systems are classified as either low temperature, medium temperature, or high temperature. Low temperature systems usually operate with a refrigerant evaporating temperature of 0°F or below. Medium temperature systems operate with a refrigerant evaporating temperature of 0° to 35°F. High temperature systems operate with a refrigerant evaporating temperature of 30° to 50°F. These systems are designed to operate with a temperature drop across the evaporator of 10°F db. Some special applications may require a different temperature drop.

## REFRIGERANT FLOW CONTROL

The most popular refrigerant flow control device used on commercial refrigeration systems is the thermostatic expansion valve (TXV). The TXV is usually set to maintain superheat settings of 8° to 12°F on high temperature systems, 5° to 8°F on medium temperature systems, and 2° to 5°F on low temperature systems.

## CONDENSING UNIT

There are generally two major components located in the condensing unit: the compressor and the condenser. Each of these components plays an important part in the operation of the refrigeration system. The heat taken from the air flowing through the evaporator coil is delivered by the compressor to the condensing coil, where it is dissipated to the cooling medium.

### *Compressor*

In effect, the compressor is the component that causes the refrigerant to circulate through the system. If the compressor is not functioning properly, the remainder of the system will not function as designed.

The compressor should be able to pump a vacuum on its low side of about 20 inches Hg. It should be capable of holding this vacuum for at least 5 minutes against a normal discharge pressure with the compressor not running. If the compressor does not meet this criteria, it will need to be replaced with one that does before full efficiency can be realized.

The suction pressure on commercial refrigeration varies according to the application and the temperature of the air flowing over the evaporator. To determine the correct suction pressure, refer to the manufacturer's specifications for the application being tested.

## Condenser

The condenser is where the refrigerant is cooled to the condensing point. If condensation is prevented, the unit will not operate satisfactorily. The condenser must be clean and free of all debris. Condensers are designed to operate with an air temperature rise of about 8° to 12°F on low temperature systems, 13° to 17°F on medium temperature systems, and 18° to 22°F on high temperature systems. If the temperature rise is out of this range, the cause must be found and corrected.

When measuring the temperature in the condenser, do not touch the coil or any metal with the thermometer or take the temperature in direct sunlight. Temperatures taken in such areas will not be accurate because of the radiant heat from the sun.

The condenser also provides the much needed subcooling of the liquid refrigerant. The desired liquid subcooling is about 10°F. The high efficiency models sometimes require a different subcooling temperature. Be sure to check the unit manufacturer's specifications for the unit make and model being serviced for the most accurate information.

The operating refrigerant discharge pressure usually corresponds to a saturation temperature of about 25° to 35°F above the ambient temperature.

Condensers are rated by the amount of total heat rejection (THR) they provide. In addition to the amount of heat removed from the cabinet or product, the heat of compression must also be removed from the refrigerant by the condenser. The compressor adds the heat of compression, which adds about 30% extra to the heat load absorbed by the evaporator. It can be calculated with the following formula:

$$CHR = cfm \times 1.08 \times \Delta T \times 0.30$$

where:  CHR = condenser heat rejection
1.08 = specific heat of air constant
$\Delta T$ = dry bulb temperature air rise through the condenser
0.30 = approximate heat of compression

To determine the cfm, use the following formula:

$$cfm = Area \times Velocity$$

### Water-Cooled Condenser

Water-cooled condensers usually require about 3 gallons per minute (gpm) per ton of refrigeration with a water temperature rise of 10°F db. Most water-cooled condensers are rated at 15,000 Btu per ton of refrigeration to include the heat of compression in the capacity, rather than the usual 12,000 Btu per ton. To determine unit capacity, use the following formula:

$$CHR = gpm \times 500 \times \Delta T$$

where:  
CHR = condenser heat rejection  
gpm = gallons per minute  
500 = a constant  
$\Delta T$ = rise in water temperature

To determine the effective refrigeration (ER), use the following formula:

$$ER = \frac{CHR}{15,000}$$

## Ice Makers

The approximate efficiency of an ice maker can be determined by weighing the amount of ice it produces in one hour of continuous operation and then dividing the harvested ice weight by the amount of heat removed from the water flowing over the ice making plate. The amount of heat removed (HR) is determined by subtracting 32°F from the entering water temperature, adding 144, then adding 0.5 Btu per °F for the difference in temperature between 32°F and the temperature of the ice when harvested. Use the following formulas:

$$HR = [(\text{Entering water temperature} - 32°F)] + 144 + [(32°F - \text{Ice temperature})(0.5)]$$

$$\text{Efficiency} = \frac{\text{lb of ice}}{HR}$$

## Determining Unit Efficiency

Use the following steps and the worksheet to determine the refrigerating capacity of the unit.

### Air-Cooled Condenser
1. Visually check the entire unit for cleanliness, and ensure all components are in proper working condition.

2. Determine the type of system (high, medium, or low temperature).
3. Set the controls to demand full refrigeration. Allow the unit to operate until the system pressures and temperatures have stabilized.
4. Install the gauge manifold on the system.
5. Measure the entering and leaving air db temperatures of the condenser.
6. Determine the condenser cfm. Use the velometer and the manufacturer's instructions.
7. Calculate the heat rejected by the condenser.
8. If this is not very close to the manufacturer's specifications, the cause must be found and corrected.

### *Water-Cooled Condenser*
1. Visually check the entire unit for cleanliness, and ensure all components are in proper working condition.
2. Set the controls to demand full refrigeration. Allow the unit to operate until the system pressures and temperatures have stabilized.
3. Install the gauge manifold.
4. Measure the entering and leaving condenser water temperatures.
5. Determine the water flow in gpm through the condenser. This can be done with a flow meter or by timing how long it takes to fill a one gallon container.
6. Calculate the heat rejected by the condenser.
7. If this is not close to the manufacturer's specifications, the cause must be found and corrected.

# Refrigeration Worksheet

Introduction: An efficiency test should be performed annually on every commercial refrigeration unit. If this procedure is followed, the equipment will operate better, use less electricity, and last longer. The technician who can perform an efficiency test properly will always be in demand. Use the following procedure along with the proper instruments to determine the efficiency of refrigeration systems.

Tools Needed: dry bulb thermometer, gauge manifold, voltmeter, ammeter, velometer, and tool kit.

Procedures:

Use the following procedure for air-cooled condensers:

1. Visually check the entire system for cleanliness, and ensure all components are in proper working condition.

2. Start the unit and allow it to operate until the system pressures and temperatures have stabilized.

3. What type of system is this (high, medium, or low temperature)? _____

4. Install the gauge manifold, and record the pressures.
   Suction _____ psig, Discharge _____ psig

5. Measure the condenser leaving air temperature (db) and record. _____ °F

6. Measure the condenser entering air temperature (db) and record. _____ °F

7. Determine the temperature rise of the condenser air. Use the following formula:

    $\Delta T$ = Leaving air temperature - Entering air temperature

    $\Delta T$ = _____ °F

8. Determine the condenser cfm. Use the following formula:

    cfm = Area x Velocity

    cfm = _____

9. Calculate the heat rejected by the condenser. Use the following formula:

    CHR = cfm x 1.08 x $\Delta T$ x 0.30

    CHR = _____ Btuh

Copyright © Business News Publishing Company

10. Is this what the manufacturer rates the equipment? _____

Use the following procedure for water-cooled condensers:

1. Visually check the entire system for cleanliness, and ensure all components are in proper working condition.

2. Start the unit and allow it to operate until the system pressures and temperatures have stabilized.

3. What type of system is this (high, medium, or low temperature)? _____

4. Install the gauge manifold, read the pressures, and record.
   Suction _____ psig, Discharge _____ psig

5. Measure the leaving condenser water temperature and record. _____ °F

6. Measure the entering condenser water temperature and record. _____ °F

7. Measure the water flow through the condenser and record. _____ gpm

8. Calculate the heat rejected by the condenser. Use the following formula:

$$CHR = gpm \times 500 \times \Delta T$$

CHR = _____ Btuh

9. Determine the effective refrigerating capacity of the equipment. Use the following formula:

$$ER = \frac{CHR}{15,000}$$

ER = _____ Btuh

10. Is this what the manufacturer rates the equipment? _____

Use the following procedure for ice makers:

1. Visually check the entire system for cleanliness, and ensure all components are in proper working condition.

2. Start the unit and allow it to operate until the first harvest cycle is complete, plus an additional hour.

3. Remove the first harvest of ice.

4. Measure the entering water temperature where it starts to flow over the plates and record. _____ °F

Copyright © Business News Publishing Company

## Operating at Peak Efficiency

5. Measure the temperature of the harvested ice and record. _____ °F

6. Weigh the total ice harvested for one hour and record. _____ lb

7. Determine the unit efficiency. Use the following formula:

$$HR = [(\text{Entering water temperature} - 32°F)] + 144 + [(32°F - \text{Ice temperature})(0.5)]$$

$$\text{Efficiency} = \frac{\text{lb of ice}}{HR}$$

Efficiency = _____ %

8. Is this what the manufacturer rates the equipment? _____

> Multiple tear-out copies
> of this worksheet
> can be found in
> Appendix B.

Copyright © Business News Publishing Company

# Chapter 8

# Megohmmeters

The proper use of a megohmmeter indicates many things to an experienced user. These instruments are used to check the very high electrical resistance of a component in millions (mega) of ohms. An example of a megohmmeter is shown in Figure 8-1. Service technicians have used them with great success on large electric motors for many years, and these uses have led to their use in many other areas of the air conditioning and refrigeration industry. However, most technicians have not had much experience with megohmmeters. This chapter presents the importance of megohmmeters in service work.

Figure 8-1. Example of a megohmmeter (Courtesy, TIF Instruments, Inc.)

Megohmmeters are currently used to check the electrical resistance of motor oil. This oil acts as an insulator in centrifugal and hermetic compressor motors. As the contaminants in the oil increase, the electrical resistance decreases. Thus, as the oil becomes contaminated, the resistance reading drops. When the contaminants are metallic particles or moisture, the resistance drops drastically, causing a much lower resistance measurement. Because of this, when regular resistance checks on a system are made and an accurate record is kept, a contaminated condition may be detected and corrected before any serious damage occurs to the system.

Because of the very high insulating effect of the winding insulation, the ordinary ohmmeter cannot detect moisture and other contaminants in the system. Ohmmeters do not generate enough voltage to detect high resistance problems that might cause problems in the very near future. On the other hand, the megohmmeter produces very high voltages, usually around 500 volts dc (vdc) and measures from 1 to 1000 ohms. An ohmmeter with a higher voltage output is not recommended for the inexperienced user, because a higher voltage may cause a weak winding to fail under test conditions, resulting in a burned out motor.

The 500 vdc capacity of the megohmmeter allows a circuit to ground to be detected if the winding insulation is weak. In this manner, a breakdown of the winding insulation can be detected and precautions taken before any real damage occurs. The megohmmeter is used to locate weak motor winding insulation and to detect moisture accumulation and acid before they have the opportunity to cause more damage. This is not to say that megohmmeters will or should replace acid test kits or moisture indicators, which are also used for this purpose.

Megohmmeters are especially useful in a preventive maintenance program. Megohmmeters are also very useful in checking refrigeration systems before signing a maintenance contract for the unit.

## Measuring Winding Resistance

The readings taken with a megohmmeter indicate the winding resistance to ground. When the insulation is in good condition, the resistance is normally very high. The 500 volt megohmmeter uses 500 vdc across the winding insulation to ground in order to measure the resistance of the insulation.

Readings should be taken immediately after the system is shut down, after operating for at least one hour. When all of the readings are taken under the same basic conditions and with virtually the same winding temperature, the readings can be properly interpreted. If any corrections are needed for various reasons, the instrument manufacturer usually provides them for the conditions encountered, such as temperature differences.

To measure the resistance, turn off all electrical power to the unit and remove the electrical wires from the terminals being tested. The megohmmeter will measure all electrical paths in the circuit being tested. When the wiring is left attached, the megohmmeter will measure the resistance of the complete circuit and will indicate the lowest resistance to ground in the circuit.

When using a megohmmeter be sure to follow the instrument manufacturer's instructions to prevent damage or electrical shock. Never use an ohmmeter on a motor winding that is under a vacuum. Be sure the instrument test leads are in proper working condition.

Equipment should be tested regularly, and these readings should be recorded. This record will indicate any changes in the condition of the system.

## SIGNIFICANCE OF THE READING

Most industry experts agree that the motor winding should have a minimum of 100 megohms resistance. Windings that have a resistance of 100 megohms to infinity are in very good condition. Windings that have less than 100 megohms are a cause for concern. Table 8-1 is a listing of the common resistance reading levels of a system with varying degrees of contamination.

| Required reading (megohms) | Condition indicated | Required preventive maintenance | Percent of windings in field |
|---|---|---|---|
| Over 100 | Excellent | None | 30% |
| 100-50 | Some moisture present | Change filter-drier | 35% |
| 50-20 | Severe moisture and/or contaminated oil | Several filter-drier changes; change oil if acid is present | 20% |
| 20-0 | Severe contamination | Check entire system and make corrections. Consider an oversized filter-drier, refrigerant and oil change, and re-evacuation. | 15% |

Table 8-1. Megohmmeter reading significance

It has been found through experience that changing the filter-drier, perhaps several times, causes an increase in the resistance. Many times this brings the resistance to above 100 megohms. When a resistance of 100 megohms or less is found, be sure to check for electrical problems in the other parts of the system, including the hermetic compressor motor terminal block.

Figure 8-2 is an example of a megohmmeter check log. You may want to make some changes to include some of the readings that are preferred for the specific installation on which you are working.

| INSULATION RESISTANCE RECORD |||||||
|---|---|---|---|---|---|---|
| Date | Meter reading ($\Omega$) | Device temperature || Air temp. (°F) | Humidity (%) | Comments |
| | | Hot | Cold | | | |
| | | | | | | |
| | | | | | | |
| | | | | | | |
| | | | | | | |
| | | | | | | |
| | | | | | | |
| | | | | | | |
| | | | | | | |
| | | | | | | |
| | | | | | | |
| | | | | | | |
| | | | | | | |
| | | | | | | |

Figure 8-2. Example of a megohmmeter check log

# Chapter 9

# Pressure-Enthalpy Diagrams

(This chapter is reprinted with permission from the booklet *Get Smart — Learn Your P-H Chart* by John Hogan, 1989.)

One diagnostic tool that is available to everyone but seldom used is the pressure-enthalpy (P-H) diagram, sometimes called the Mollier diagram. This tool is used by most design engineers, but unfortunately it is overlooked by service technicians.

The important thing to remember about the P-H diagram is that it cannot be used to plot every system that is serviced. However, just learning the chart provides a new and fresh insight on what is happening inside the system.

The P-H diagram is a simple, graphic way to plot a system cycle and observe its characteristics. No two systems are exactly alike. Each one has its own personality. When using the diagram, a single pound of refrigerant is followed completely around the system in order to learn when and where it changes states and what those changes mean. Since only one pound of refrigerant is followed, the diagram may be applied to any size system. The pounds per minute circulated will determine the total system capacity. To better understand this process, some basics must be reviewed, starting with diagram construction.

## SATURATION, SUBCOOLING, AND SUPERHEAT

In order to become a proficient technician, you must learn about the three conditions in which a refrigerant can exist while it is traveling around the system. These conditions are: saturation, subcooling, and superheat. To pinpoint which condition the refrigerant is in at any given point, both a temperature reading and a pressure reading must be taken. Taking only one of these readings will not be sufficient.

A good method for learning about the three conditions is to study water, which is something most people are familiar with. Water and refrigerant are similar, except that at atmospheric pressure (0 psig) water boils at 212°F, whereas R-22, for example, boils at -41°F at the same pressure. The boiling point is called the *saturation temperature*.

The diagram in Figure 9-1, plots the heat content of water in Btu/lb along the horizontal axis versus the temperature in °F along the vertical axis. There are two types of heat involved in the process depicted in Figure 9-1: *sensible heat* and *latent heat*. Sensible heat involves a change in temperature, and latent heat involves a change of state (at a constant temperature). The diagram is plotted at 0 psig (atmospheric) pressure. As the water is warmed from 52°F to 212°F, only sensible heat is used. This is because the only change experienced is a change in temperature. Since subcooling is defined as the number of degrees the liquid temperature is below its saturation temperature, the liquid is in the subcooling state. At 52°F, the subcooling is 160°F (212°F - 160°F = 52°F). At 212°F, the subcooling is 0°, because it has reached the saturation point.

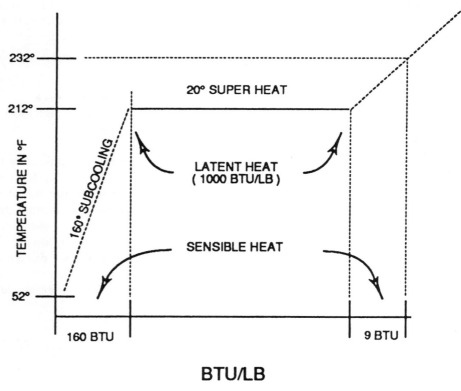

Figure 9-1. Sensible and latent heat

When the water reaches its saturation point of 212°F at atmospheric pressure, boiling will start at a constant temperature as long as the pressure remains constant. During this change of state from a liquid to a vapor, the water absorbs approximately 1000 Btu for each pound of water evaporated. This is called the *latent heat of evaporation*.

When all of the liquid has boiled off and more heat is added to the vapor, the vapor temperature rises above the saturation temperature of 212°F. Since the temperature of the vapor changes, sensible heat is again experienced. The water now exists as a vapor and is in the superheated state. Superheat is defined as the amount of degrees the vapor temperature is above the saturation temperature at a given pressure. If the temperature of the superheated vapor is measured at 232°F, then the amount of superheat would be 20°F, because it is 20°F above the saturation temperature of 212°F.

Note the tremendous amount of heat required to boil a fluid or cause it to go through a change of state (latent heat of evaporation) when compared to the amount of heat it takes to simply change its temperature. For this reason, the evaporator should be kept as full of liquid refrigerant as possible without flooding back to the compressor in order to get the maximum amount of heat transfer.

Since all of the conditions shown on the diagram occur at one pressure (0 psig), it is obvious that just taking a pressure reading will not indicate what the system is doing. In order to determine if the refrigerant is subcooled, saturated, or superheated, the temperature at a given point must also be known. Whenever the pressure reading is taken, immediately check a pressure-temperature (p-t) chart to see what the saturation temperature is for the refrigerant at that pressure. When the temperature at that point is known, it can then be determined if the refrigerant is subcooled, saturated, or superheated. If the temperature is above the saturation temperature, then it is superheated. If the temperature is below the saturation temperature, then it is subcooled. If the temperature is the same as the p-t chart indicates, then both saturated liquid and saturated vapor are present.

## THE THREE ZONES

The P-H diagram is divided into three zones: saturated, subcooled, and superheated, Figure 9-2. As the refrigerant travels around the system, it exists in one of the three zones indicated on the diagram. The refrigerant is always changing conditions and never exists in any two zones at the same time. The refrigerant is saturated, subcooled, or superheated. If technicians do not have an understanding of these basic principles, they will never thoroughly understand how refrigeration systems operate.

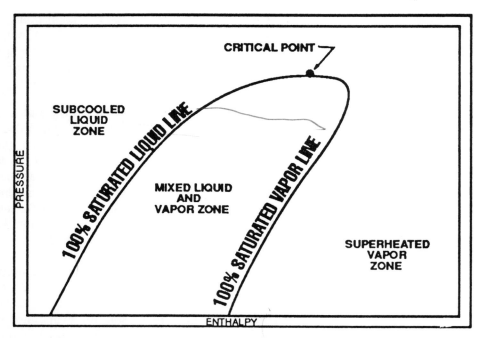

Figure 9-2. P-H diagram zones

### Saturated Zone

Any time the refrigerant is inside the envelope or dome, it is in the *saturated zone*. When a refrigerant is saturated, it contains both liquid and vapor. The right-hand curve of the saturated zone is the *100% saturated vapor line*, and the curve on the left-hand side of the zone is the *100% saturated liquid line*. Some P-H charts have a series of vertical lines within the saturated zone. These lines indicate the percentage of vapor along each line. Saturation occurs in the evaporator, where the refrigerant changes state from a liquid to a vapor (boiling), and in the condenser, where the refrigerant changes state from a vapor to a liquid (condensing). As previously noted, the temperatures and pressures shown on the pressure-temperature chart are all in the saturated state.

### Superheated Zone

To the right of the 100% saturated vapor line is the *superheated vapor zone*, where the refrigerant is above the saturation temperature. The superheated condition should exist from within the outlet of the evaporator to within the inlet portion of the condenser. The superheat measurement is the temperature of the vapor minus the saturation temperature at a given pressure.

## Subcooled Zone

To the left of the 100% saturated liquid line is the *subcooled liquid zone*. The refrigerant is in the liquid state, and its temperature is below the saturated temperature at a given pressure. The subcooled condition should exist from the outlet of the condenser to the inlet of the expansion device. The amount of subcooling is found by subtracting the temperature of the liquid from the condensing temperature at a given pressure.

## Critical Point

The critical point on the P-H diagram is located where the saturated liquid curve and the saturated vapor curve converge. At any temperature above this point, the refrigerant may exist in the vapor condition only.

# Five Refrigerant Properties

The following five refrigerant properties are graphically depicted in Figures 9-3 through 9-7: absolute pressure, enthalpy, constant temperature, entropy, and constant volume.

Most refrigeration gauges read in pounds per square inch gauge (psig) above atmospheric pressure. At sea level, or 0 pounds gauge, the pressure would be 14.7 pounds of atmospheric pressure. Pressures below 0 psig are considered a partial vacuum and read in inches of mercury (Hg). A perfect vacuum is defined as 0 pounds per square inch absolute (psia). The P-H diagram in Figure 9-3 is scaled for *absolute pressure*.

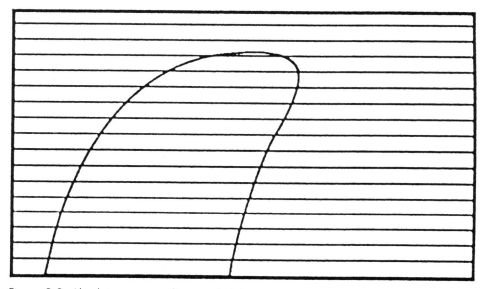

Figure 9-3. Absolute pressure diagram (psia)

Most computations use psia. To obtain absolute pressure simply add 14.7 psi to the gauge reading in psig. For example, to convert a gauge pressure reading of 10 psig to psia, add 10 psig + 14.7 psi. This gives an absolute pressure reading of 24.7 psia. Many people forget to make this conversion when reading the P-H diagram. Always convert pressures to absolute for this use.

*Enthalpy* represents the total heat content in Btu per pound. The enthalpy scale shown in Figure 9-4 is usually shown at both the top and bottom of the diagram, and the lines of constant enthalpy run vertically.

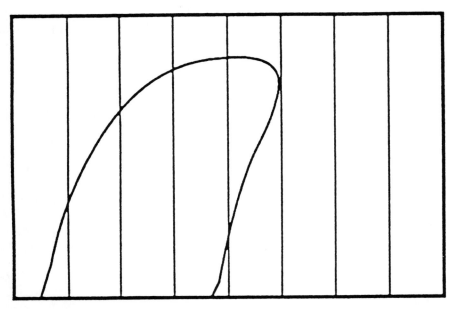

Figure 9-4. Enthalpy diagram (Btu/lb)

In Figure 9-5, please note that the lines of *constant temperature* run almost vertically in the superheated zone, horizontally in the saturated zone, and vertically in the subcooled zone.

*Entropy* is defined as the heat available measured in Btu/lb per degree of change for a substance. The lines of constant entropy extend diagonally up to the right from the saturated vapor line, Figure 9-6. Entropy is used mainly for engineering calculations. The main concern of the service technician is that when the refrigerant is compressed, it follows up the line of constant entropy.

The lines of *constant volume* extend out from the saturated vapor line into the superheated zone, Figure 9-7. These values indicate how much space (cu ft) is taken up by each pound of refrigerant vapor.

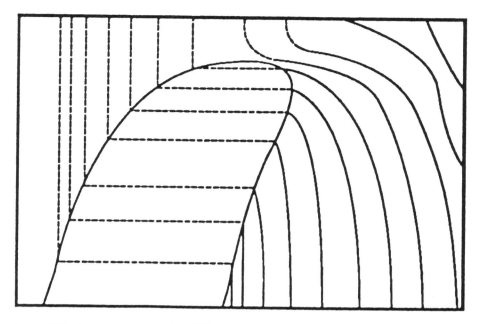

Figure 9-5. Constant temperature (°F)

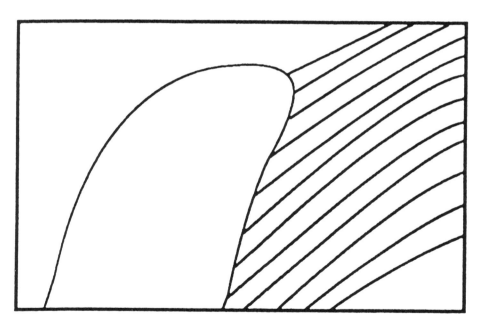

Figure 9-6. Entropy

## THE IDEAL CYCLE

Usually, the refrigeration cycle is shown as an *ideal cycle*. This is generally an illustration of a cycle, but it does not show superheating in the evaporator, suction line, heat exchangers, vapor-cooled compressors, etc. Also, subcooling of the liquid and pressure losses are not shown.

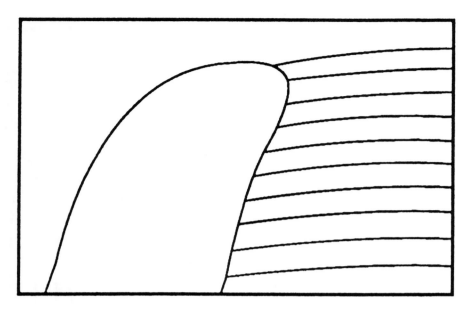

Figure 9-7. Constant volume (cu ft/lb)

These things are important to the service technician, but to understand the basic cycle, the ideal cycle will be discussed first.

The four functions of the refrigeration cycle are: compressing, condensing, expanding, and evaporating. The following is a discussion of these four functions.

To draw a simple ideal cycle, only two pressures are required: the evaporation pressure and the condensing pressure. They are also called the *high side* and *low side* pressures. Do not forget to add 14.7 psi to the gauge readings to obtain the absolute pressure required for the diagram. Figure 9-8 shows the condensing and evaporating horizontal pressure lines.

Next, drop the expansion line down from the intersection of the condensation line and the saturated liquid line to intersect with the horizontal evaporation line, Figures 9-9 and 9-10.

Finally, complete the cycle by extending the compression line up to the superheated zone, parallel with the constant entropy lines, until it intersects the condensation line, Figure 9-11.

Figure 9-12 shows a pound of refrigerant as it travels around the ideal cycle. The refrigerant leaves the compressor as high pressure vapor in a highly superheated condition (line a to b). The refrigerant vapor is compressed and thereby condensed into a liquid with such cooling media as air or water. It is then reused for another cycle. As the pound of refrigerant moves left, horizontally along the condensing pressure line, the condenser must remove the superheat before it reaches the 100% saturated vapor line (line b to c).

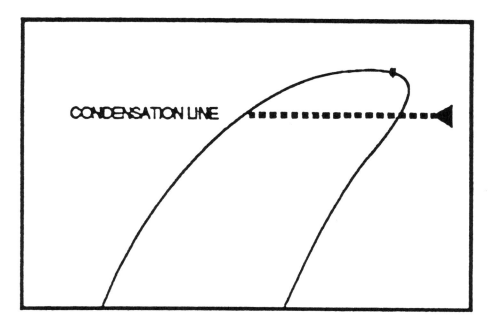

Figure 9-8. Condensation line

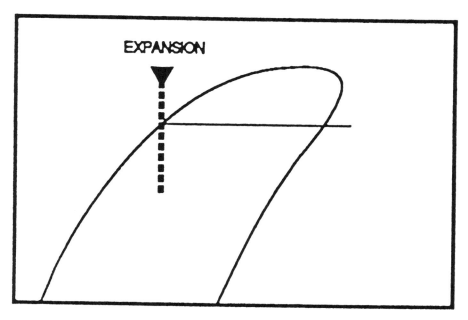

Figure 9-9. Expansion line

# Operating at Peak Efficiency

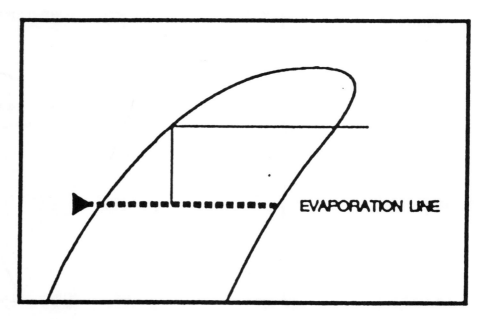

Figure 9-10. Evaporation line

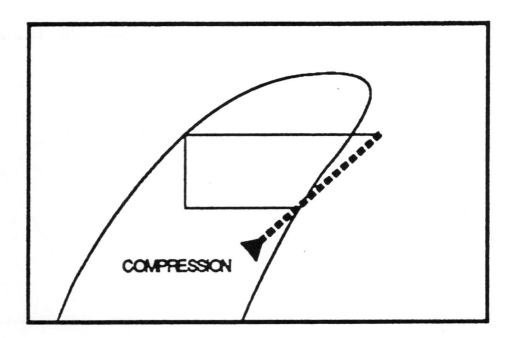

Figure 9-11. Compression line

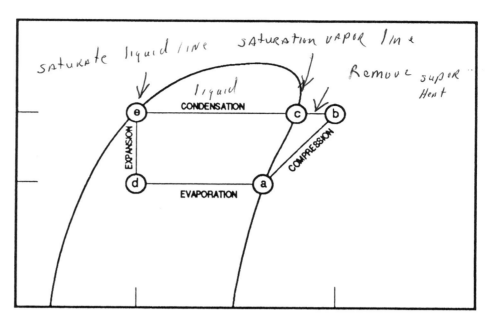

Figure 9-12. The ideal cycle

Condensation starts when the refrigerant crosses the saturated vapor line (point c) into the saturated zone. Note that as the refrigerant moves along the constant pressure line in the saturated zone, the temperature remains constant. When the pound of refrigerant reaches the 100% saturated liquid line (point e), all of the vapor has condensed to a liquid.

The next step is the expansion process, during which the pressure is dropped suddenly from high pressure to low pressure within the expansion device (and refrigerant distributor and tubes if used). There is no external transfer of heat into or out of the refrigerant in the ideal cycle process; therefore, the refrigerant flows vertically down the line of constant enthalpy (line e to d). Since it is crossing the horizontal lines of temperature while in the saturated zone, there is a temperature drop in the mixture. This is due to the latent heat of vaporization as a portion of the refrigerant flashes into a vapor. The refrigerant is now ready to enter the evaporator (point d).

The refrigerant enters the evaporator at the evaporator pressure in the saturated zone. It passes along the constant pressure line in the saturated zone to the 100% saturated vapor line (point a). As previously mentioned, some P-H diagrams have vertical lines in the saturated zone which represent the percentage of vapor in the mixture. As the refrigerant passes through the evaporator, it gains heat as it boils from a liquid to a vapor. When the refrigerant reaches the 100% saturated vapor line, all the heat absorbed is contained in the vapor. It is now ready to be compressed again.

Do not forget, this is just the ideal cycle. A more realistic cycle will be discussed later in the chapter.

## PERFORMANCE OF THE CYCLE

With the help of the P-H diagram, certain quantities of the cycle may be obtained. The engineer who wants more exact figures should use the various refrigerant tables. The P-H diagram, remember, is made up from the values in these tables. However, it is much simpler to use the P-H diagram.

The quantities that can be obtained are the heat of rejection, the refrigerating effect, the circulation rate, the compression ratio, the work of compression, the coefficient of performance, the volume rate of flow per ton, and the power per ton. The purpose of this presentation is to help the technician better understand the refrigeration cycle through the study of the P-H diagram and not to confuse things with a lot of mathematical formulas; therefore, we will just discuss the first four points mentioned above. For the clarification of these points, we will consider an R-22 system that develops 10 tons of refrigeration while operating with a 90°F condensing temperature and 20°F evaporating temperature.

### Heat of Rejection

The *heat of rejection* is the heat given up by the condenser, Figure 9-13. If the one pound of refrigerant has an enthalpy (total heat content) entering the condenser of 118 Btu/lb (h3) and an enthalpy leaving the condenser of 38 Btu/lb (h1), the difference is the heat given up by the condenser. Note that the condenser is rejecting the heat picked up by the evaporator plus the heat of compression. The *heat of compression* is the heat added to the refrigerant by the work done by the compressor. Use the following formula to find the heat of rejection:

Heat of rejection = h3 - h1 = 118 Btu/lb - 38 Btu/lb = 80 Btu/lb

where: h = enthalpy

### Refrigerating Effect

*Refrigerating effect* is the heat absorbed by the evaporator, as indicated by h2 and h1 in Figure 9-14. Since the total heat content is 38 Btu/lb entering and 107 Btu/lb leaving, then the heat picked up in the evaporator is:

Refrigerating effect = h2 - h1 = 107 Btu/lb - 38 Btu/lb = 69 Btu/lb

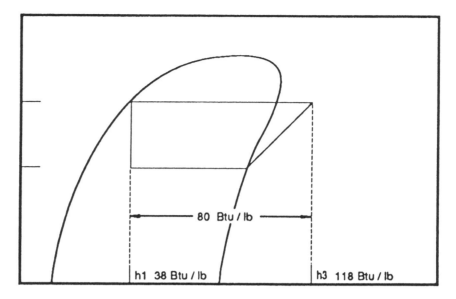

Figure 9-13. Heat of rejection

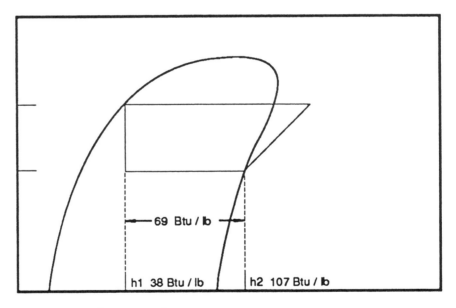

Figure 9-14. Refrigerating effect

## Circulation Rate

If the refrigerating capacity is known, we can then determine the total *circulation rate* in lb/min by dividing the capacity in Btu/min by the refrigerating effect of 69 Btu/lb.

$$\text{Refrigerant flow} = \frac{10 \text{ tons} \times 200 \text{ Btu/min/ton}}{69 \text{ Btu/lb}} = 28.98 \text{ lb/min}$$

### Compression Ratio

The *compression ratio* is the absolute discharge pressure divided by the absolute suction pressure, Figure 9-15. Consequently, the higher the suction pressure the lower the compression ratio at a given discharge pressure. Also, the lower the discharge pressure the lower the compression ratio.

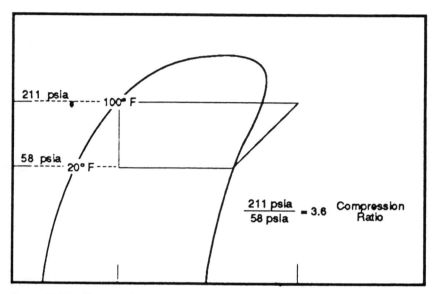

Figure 9-15. Compression ratio

While maintaining operating temperatures to satisfy the product requirements, try to keep the compression ratio as low as possible. The lower the compression ratio the higher the volumetric efficiency. The higher the volumetric efficiency the more mass flow of refrigerant pumped by the compressor. The more mass flow pumped by the compressor the greater the system capacity, and less running time is required. The less running time required, the more money saved.

As strange as it may seem, keeping the suction pressure up has a greater effect in lowering the compression ratio than lowering the discharge pressure by an equal amount. An added benefit of a lower compression ratio is lower discharge temperatures, which in turn result in less refrigerant-oil breakdown and less contaminants formed by high operating temperatures.

## THE ACTUAL REFRIGERATION CYCLE

Remember that the ideal cycle discussed earlier in the chapter did not have pressure drops, subcooling, or evaporator superheat. Such a system does not exist except on paper. Without evaporator superheat in the refrigeration cycle, there would be liquid slugging the compressor; without pressure drops, there would be no refrigerant flow. However, the

# Pressure-Enthalpy Diagrams

ideal cycle is an excellent place to start when attempting to understand the basic P-H diagram and the refrigeration cycle.

To begin, consider a simple system with the following basic components: evaporator, compressor, and expansion device. The other components can be added later. This section will illustrate some of the horrors that can happen in the field and show how your knowledge of the P-H diagram can be used to provide a solution to these problems.

## Low Pressure Line

Certain assumptions have to be made to illustrate the difference between the ideal cycle and the actual cycle. Let's start with the low pressure line. Notice that it is no longer called the evaporator line, because it now involves more than just the evaporator. There is a pressure drop through the evaporator tubes and suction line, superheat is developed in the evaporator, and additional superheat occurs in the suction line. Also, assume that this is a semi-hermetic, vapor-cooled type compressor. There is also additional superheat in the compressor body before the compression begins.

As shown in Figure 9-16, there is quite a difference in the discharge temperature between the two plots. Also, notice the "spike" in the actual cycle discharge temperature at the peak of its plot. This illustrates the high temperature experienced at the discharge valve of the compressor. Compressor engineers indicate that this temperature is approximately 75°F higher than the temperature measured about 6" down the discharge line from the discharge service valve. Consequently, if the oil-refrigerant mixture starts to breakdown at about 300°F, then the temperature of the discharge vapor should be limited to about 225°F.

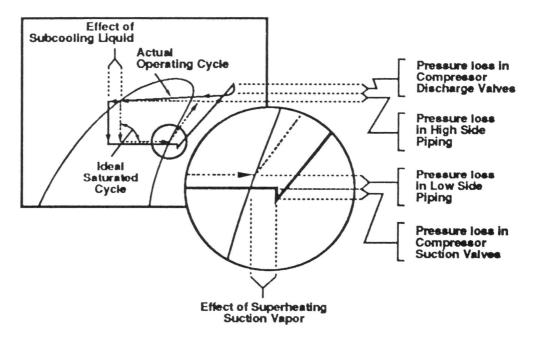

Figure 9-16. Plotting the actual refrigeration cycle on a P-H diagram

By observing the diagram, it is evident that moving the point where compression starts toward the left decreases the chances of running excessively high discharge temperatures, which result in refrigerant-oil breakdown. The following are some ideas to reduce the discharge temperature:

1. Check to see if the evaporator superheat is too high.
2. Re-route the suction line if possible to a cooler location.
3. Check to see if a liquid-suction heat exchanger is used unnecessarily.
4. Inject liquid into the suction line (consult the TXV manufacturer).

The idea behind all of these steps is to move the compression line further to the left by decreasing the superheat and the discharge gas temperature. Also, if the superheat in the evaporator is too high, lowering it to the recommended amount increases the evaporator capacity, because it contains more refrigerant. This also raises the suction pressure, which results in a lower compression ratio, which in turn increases system capacity.

Some people equate frost with liquid refrigerant presence. **Do not adjust the superheat to a frost line.** Admittedly, there could possibly be some liquid present when frost is present, but not always. It depends on the operating pressures and temperatures of the system.

Suppose we had an ice cream case operating at a -40°F evaporator saturation temperature, and we adjusted to a frost line at the suction service valve. This means the temperature at the service valve would be about 32°F. Since superheat is defined as the vapor temperature minus the evaporator saturation temperature, this would result in a 72°F superheat at the suction service valve (32°F - -42°F = 72°F). This is excessive. Generally, case manufacturers recommend a 7°F superheat in the evaporator. With the valve adjusted properly and assuming a 20°F rise in the suction line, the temperature at the suction service valve would be -13°F. There would certainly be frost at this temperature. However, there would be no liquid refrigerant present. This type of adjustment also makes quite a difference in the operating efficiency of the system.

### High Pressure Line

The high pressure line consists of the discharge line, condenser, receiver (if used), and the liquid line. Since the refrigerant is still superheated when it enters the condenser and is subcooled when it leaves the condenser, the condenser is the only component in which all three states (superheat, saturation, and subcooling) exist simultaneously. There are some systems that have the liquid line subcooled below the evaporator saturation temperature. This would produce all three states in the evaporator, but these systems are the exception rather than the rule.

# PRESSURE-ENTHALPY DIAGRAMS

If a liquid-to-suction heat exchanger is used, about 10°F of subcooling is possible, and the vertical expansion line will move to the left. Now there is a smaller value of enthalpy entering the evaporator, which produces a greater capacity. Remember that while adding a heat exchanger increases the capacity on the low side, it causes the opposite effect on the high side of the system. About 15°F can be added to the suction gas temperature entering the compressor, which, as we discussed, results in a higher discharge temperature. Before arbitrarily adding a heat exchanger, determine that this is a step that needs to be made. Sometimes a heat exchanger is required to subcool the liquid to ensure 100% liquid at the expansion device inlet. This would justify the use of a heat exchanger.

Suppose the subcooling at the receiver is only 2°F, and there is a 10 ft lift in the liquid line to the expansion device. There would be a minimum 5 psi pressure drop, because there is a 1/2 lb drop for every foot of lift. Add to this the line loss drop due to friction, plus the drop through the accessories. As shown in Figure 9-16, the pressure drop could carry the refrigerant down into the saturation zone where vapor would exist. The challenge is to move the vertical line to the left by subcooling or to elevate the horizontal pressure line so that even with the pressure drops there will still be subcooled liquid in the subcooled zone.

The following are some suggested methods to increase the subcooling and move the vertical line to the left:

1. Liquid-to-suction line heat exchanger
2. Liquid-to-expansion device outlet heat exchanger
3. Mechanical subcooling
4. Water-cooled subcooler
5. Surge receiver

## Expansion Line

Most P-H diagrams show the expansion line as the pressure drop only through the expansion device. This is true is some cases, but over the years there has been a tremendous increase in the use of refrigerant distributors. For this reason, a distributor has been added to the example. The drop from the thermostatic expansion valve (TXV) inlet to the evaporator inlet includes both the TXV and distributor. If a system were operating under full load (200 psig at the TXV inlet and 70 psig at the evaporator inlet), the total pressure drop would be 130 psig. However, the drop across the TXV would be about 100 psig, and the distributor tubes would be approximately 30 psig. This must be considered when selecting the TXV.

Today, mechanical subcooling is a design concept used by many engineers. As shown in Figure 9-16, with an increase in subcooling, the pressure drops lower before the expansion line crosses into the saturated zone. This means that not only is system capacity increased because of lowering the entering heat content, or enthalpy, but a greater mass flow through the expansion device is experienced, because the vapor is formed during the expansion process. This is important to know when sizing the expansion device. Many systems are operating with oversized TXVs because of a high degree of subcooling. This sometimes results in liquid floodback to the compressor. Expansion valve manufacturers provide subcooling capacity factors in their catalogs. Remember that subcooling also increases distributor and tube capacities and should be checked for proper sizing. Quite often, extensive mechanical subcooling is used on a system although the TXV and distributor sizing have not been considered. This results in poor system performance and poor refrigerant distribution and floodback.

### *Plotting The EPR Valve*

The evaporator pressure regulating (EPR) valve can be easily plotted on the P-H diagram. If the system has a single evaporator and a single compressor, there would not be much change in the plot, because they are selected at a low pressure drop of 2 psi or less. This small amount would not show much on the diagram. However, if there is a supermarket system that has a high temperature evaporator, then there could be as much as a 20 psi or more pressure drop across the valve. The reason for mentioning the EPR is that many people expect a high temperature drop to accompany the pressure drop.

If the EPR valve is located in the machine room and there is a superheat of about 40°F at the EPR location, then the pressure would be dropped down 20 psi from this point. Notice that since the plot is in the superheated zone, the temperature lines run almost vertically and thus only a slight temperature change would be detected, if any. If the operation was in the saturation zone then there would be a dramatic temperature drop.

The P-H diagrams that follow are used by permission of DuPont Fluoroproducts.

# PRESSURE-ENTHALPY DIAGRAMS

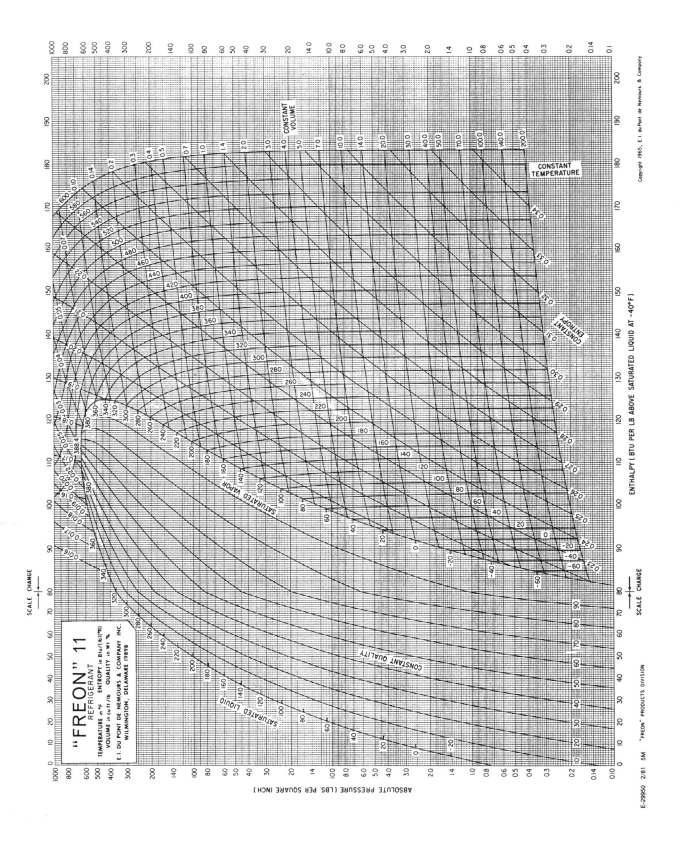

# Operating at Peak Efficiency

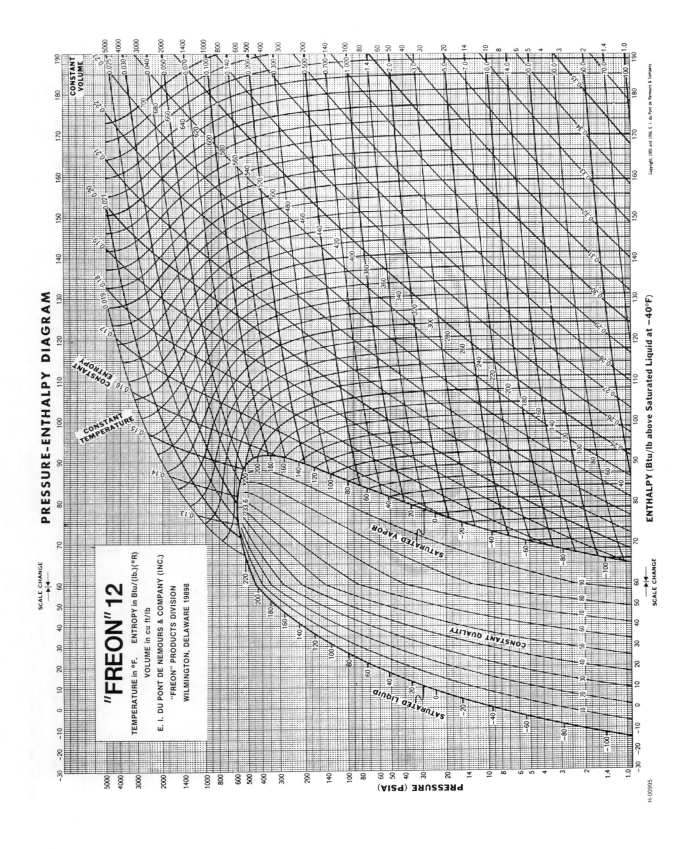

# PRESSURE-ENTHALPY DIAGRAMS

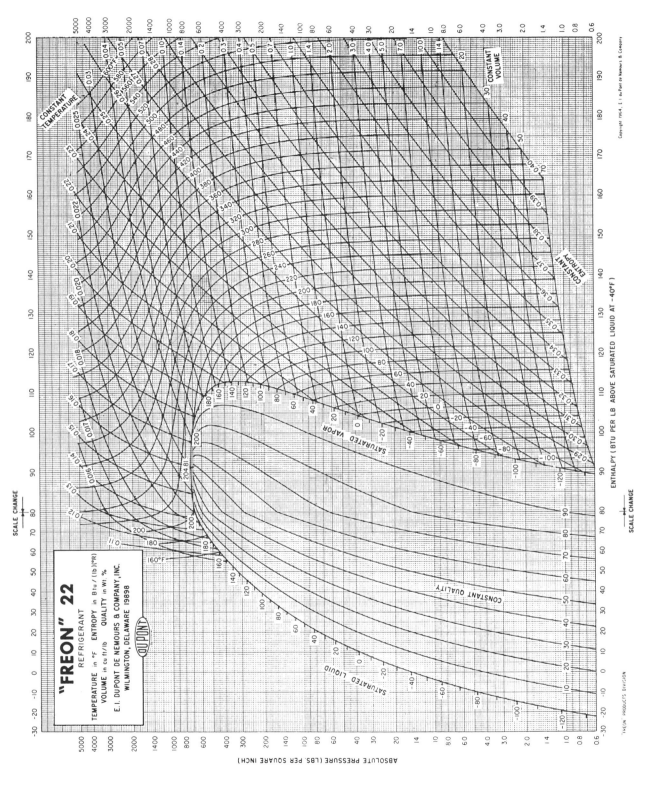

# Operating at Peak Efficiency

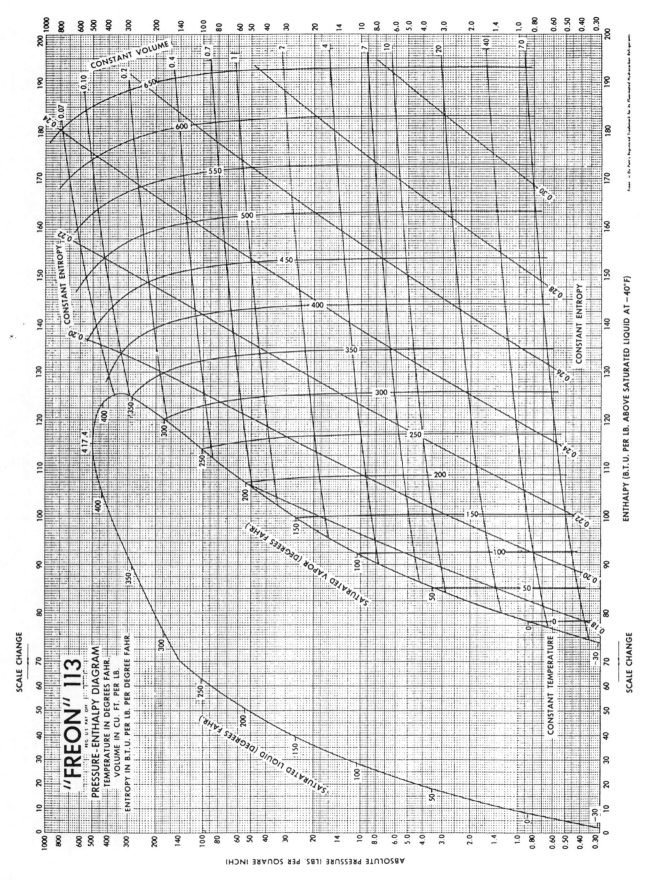

# Pressure-Enthalpy Diagrams

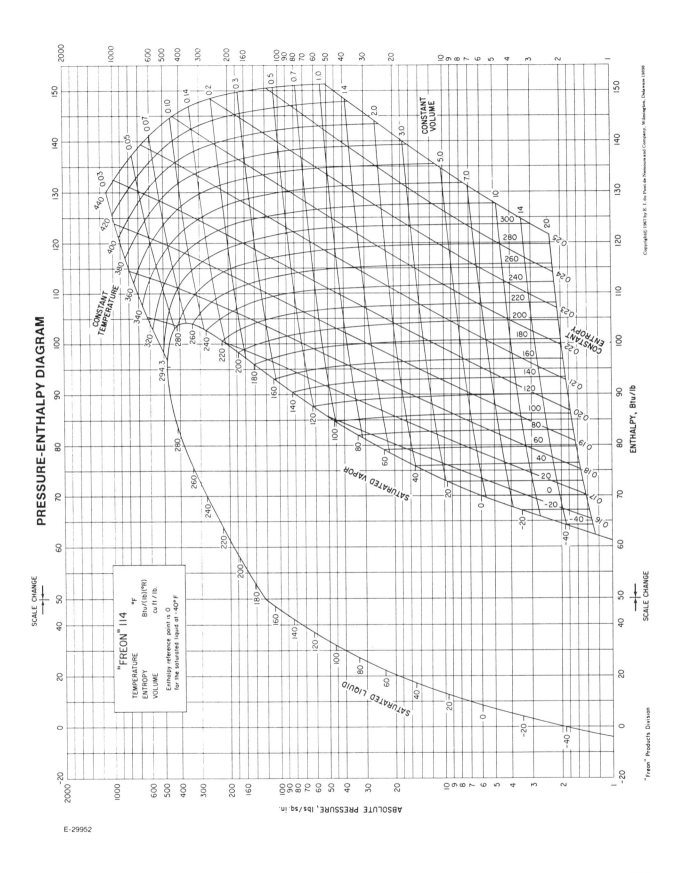

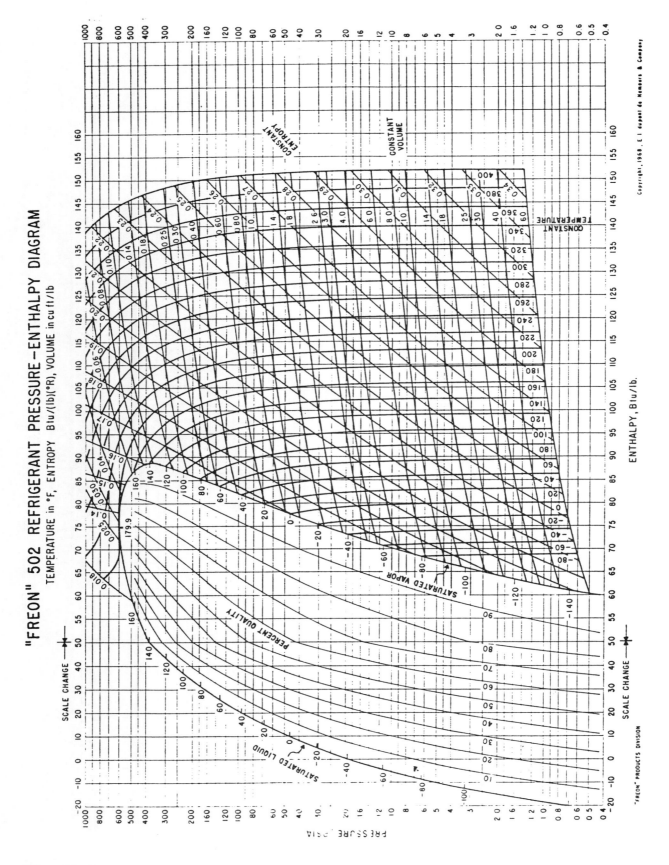

# Pressure-Enthalpy Diagrams

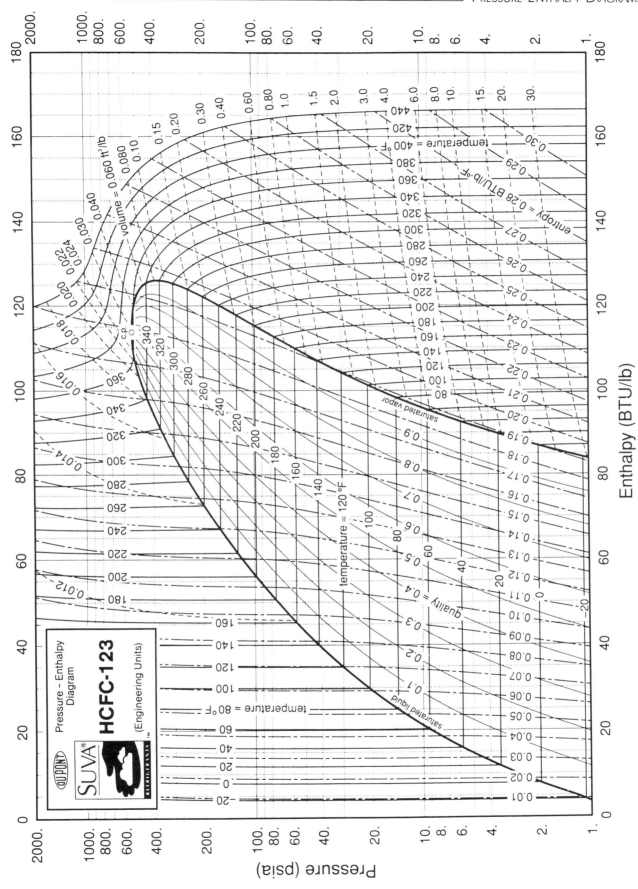

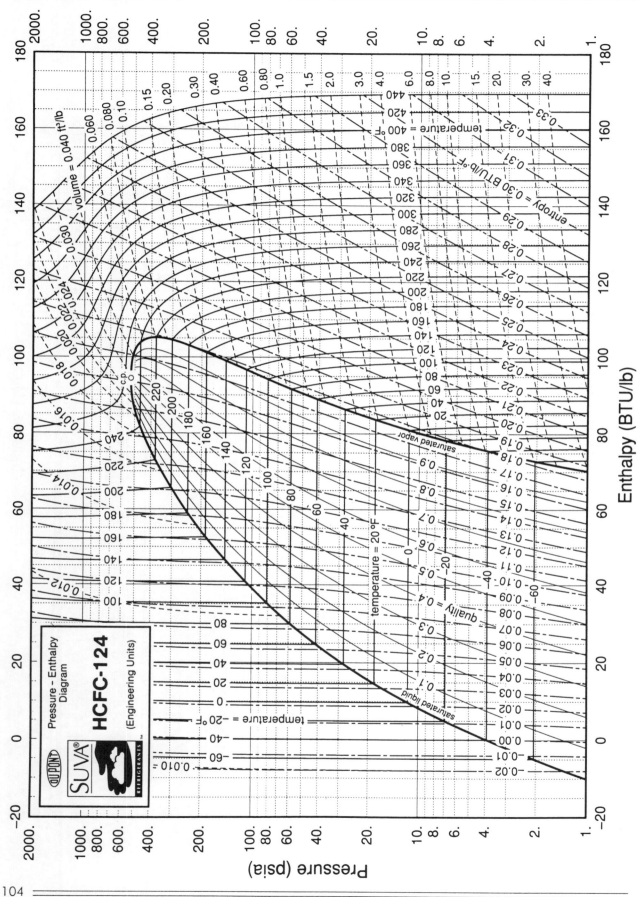

# Pressure-Enthalpy Diagrams

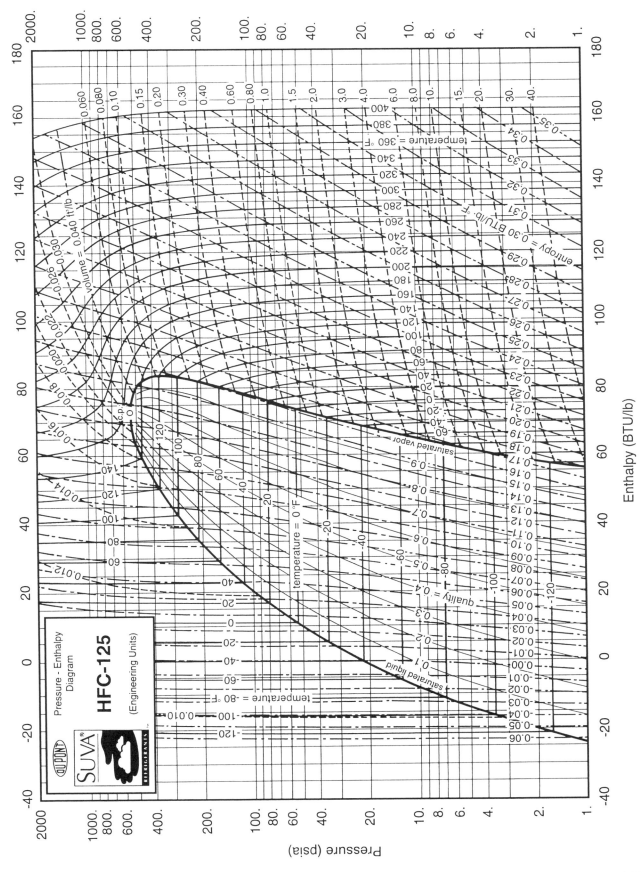

## Operating at Peak Efficiency

# Pressure-Enthalpy Diagrams

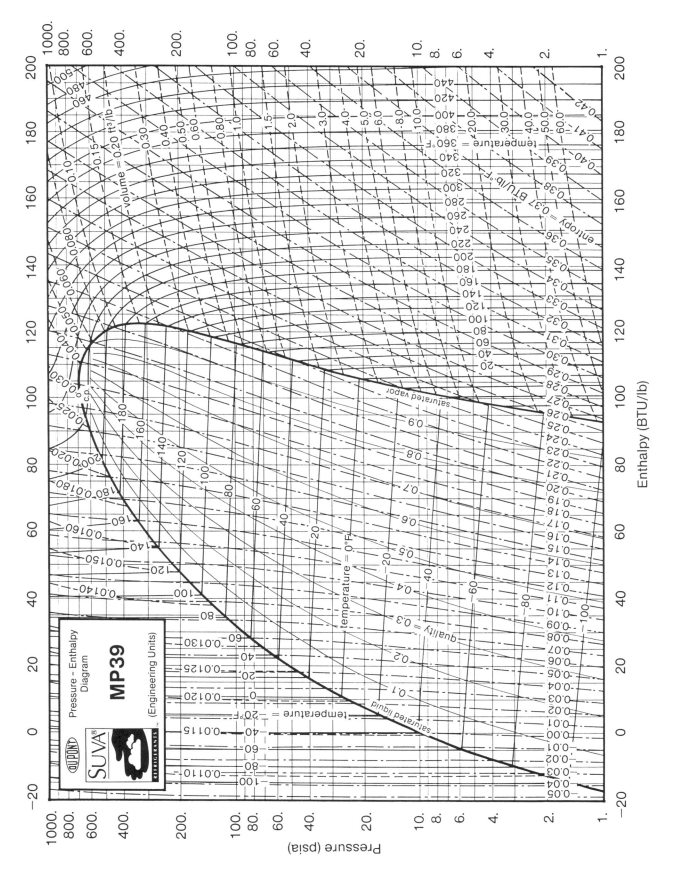

# Operating at Peak Efficiency

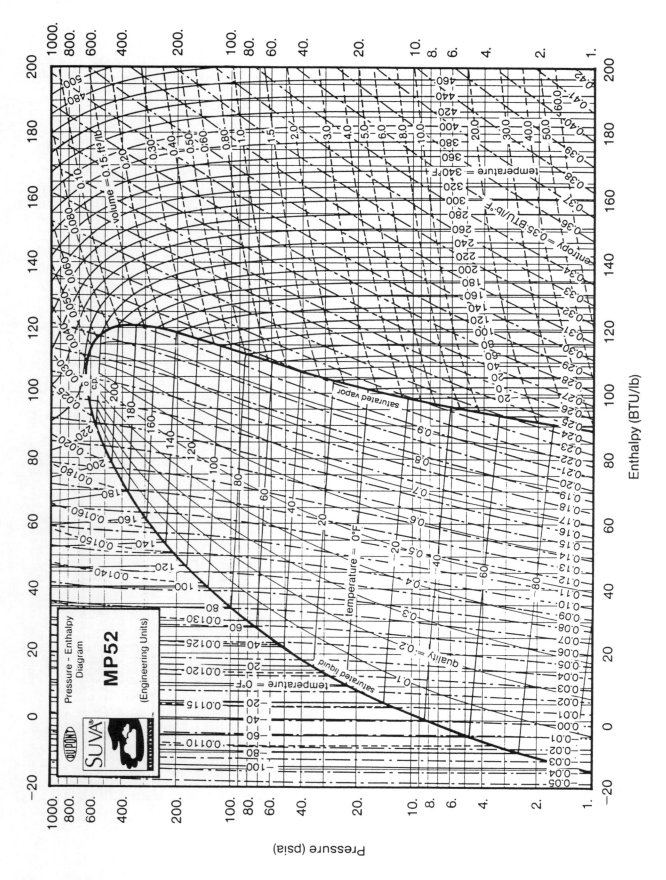

# Pressure-Enthalpy Diagrams

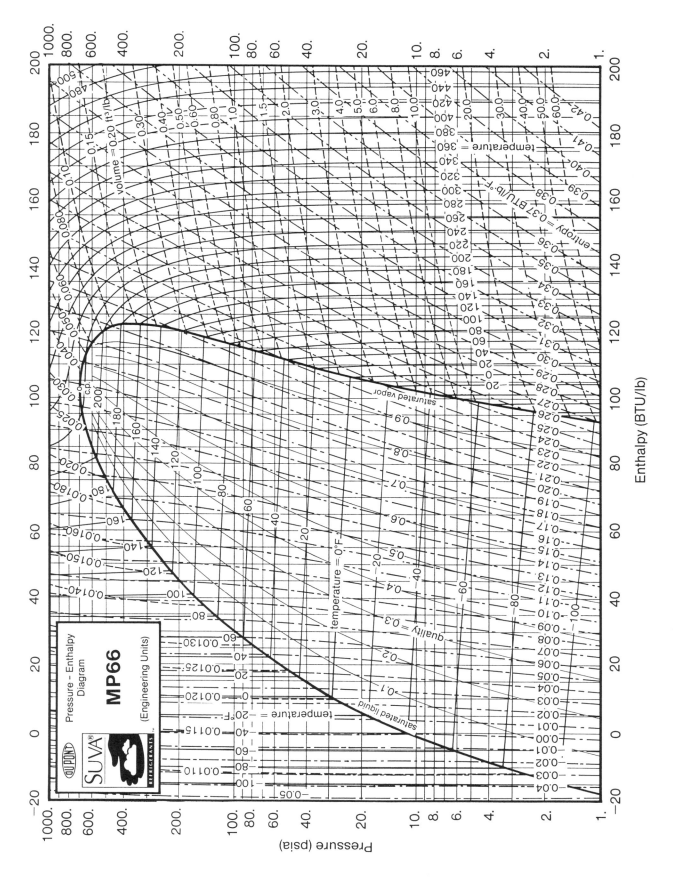

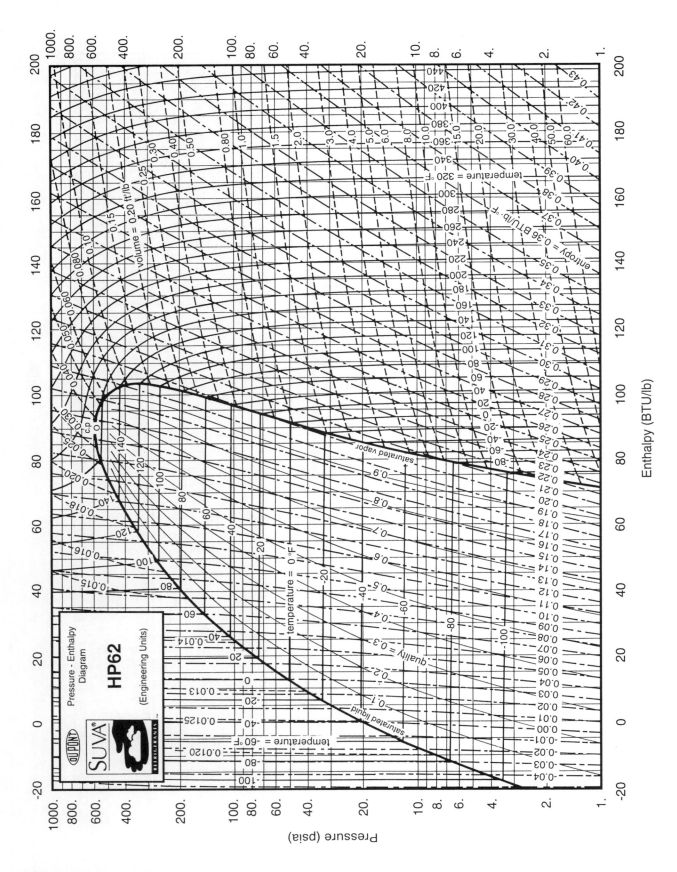

## Pressure-Enthalpy Diagrams

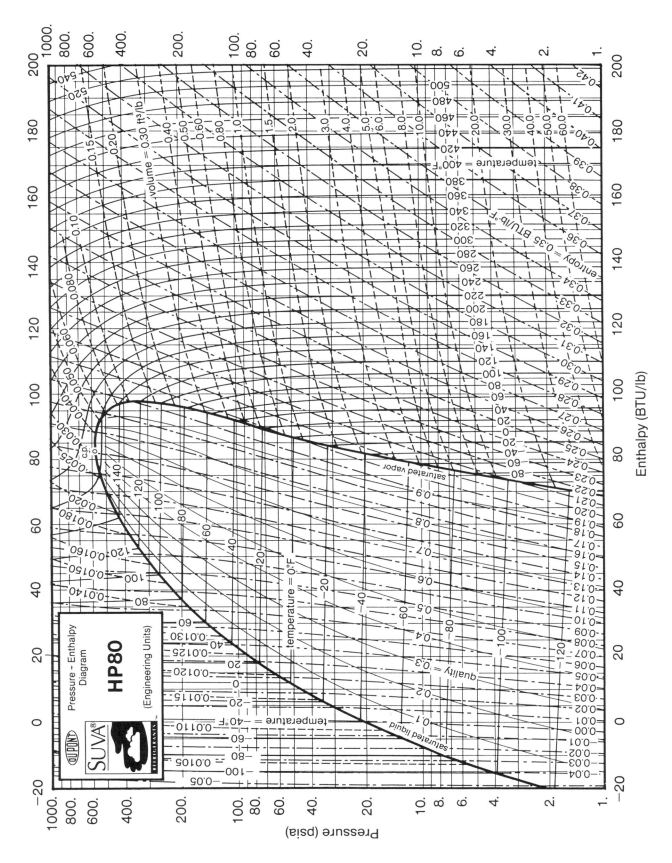

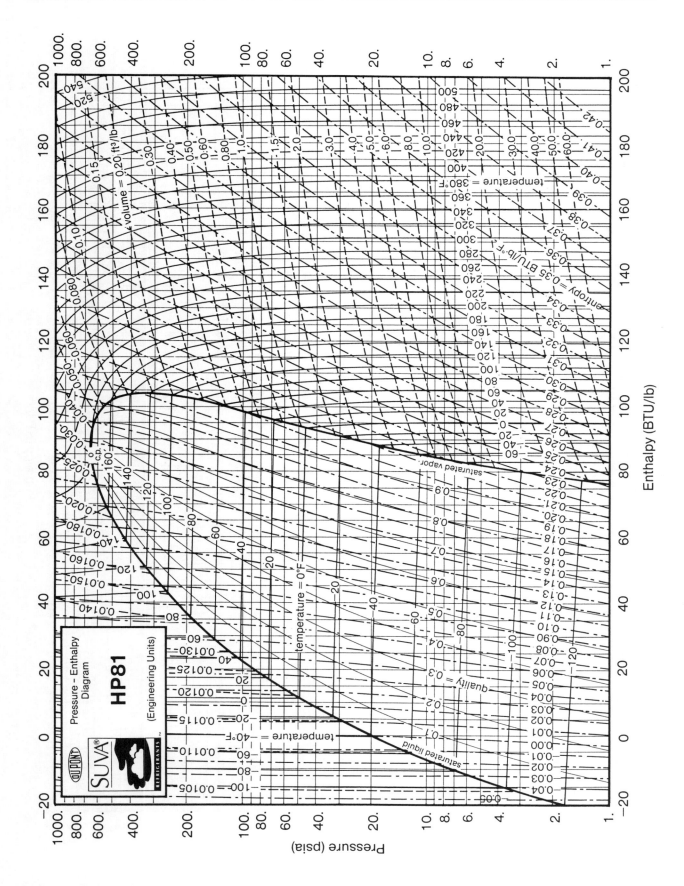

# Chapter 10

# Estimating Annual Operating Costs

Technicians and engineers are often called upon to determine how much a central air conditioner or heat pump will cost to operate for a season, or in other words, the annual operating cost. To help make this determination, the Air-Conditioning and Refrigeration Institute (ARI) has developed an energy guide and the *ARI Directory of Certified Unitary Air-Conditioners, Unitary Air Source Heat Pumps, and Sound-Rated Outdoor Unitary Equipment*. This directory lists the information required from each manufacturer to properly complete the annual operating cost estimate.

After checking the directory, simply locate the equipment manufacturer by unit and model number, place the appropriate figures in the worksheet spaces, and make the required calculations to provide the estimated annual operating costs to the customer. There is a sample problem provided in this chapter, which guides the estimator through the required steps. The estimator should check with the local power supplier to determine the actual cost of electricity so that the operating cost estimate is more accurate.

The following worksheets are reprinted courtesy of Air-Conditioning and Refrigeration Institute (ARI), February 1, 1994 - July 31, 1994, *Directory of Certified Unitary Equipment*.

OPERATING AT PEAK EFFICIENCY

ARI GUIDE FOR
ESTIMATING ANNUAL OPERATING COST
OF A CENTRAL
AIR CONDITIONER OR HEAT PUMP

This guide is designed to assist consumers in estimating their annual operating costs of central air conditioners and heat pumps covered by the U.S. Federal Trade Commission (FTC) appliance labeling rules. Along with the energy efficiency information contained in this directory, this guide is designed to assist consumers in making purchasing decisions. Contained in the guide are step-by-step instructions on how to perform the operating cost estimates. The operating cost estimates can be performed using data included in this directory. A sample worksheet is also provided.

Note. In the following text Energy Efficiency is expressed in 3 ways:

SEER, Seasonal Energy Efficiency Ratio (for cooling).
HSPF, Heating Seasonal Performance Factor (for heating).
EER, Energy Efficiency Rating (a term used by the Federal Trade Commission to mean <u>either</u> SEER or HSPF, whichever is applicable).

To assist consumers in making informed decisions regarding equipment selection, the FTC has determined that the minimum and maximum product energy efficiency ratings available are those listed below:

| DESCRIPTION | | ENERGY EFFICIENCY RATING (EER) RANGES | | | |
| --- | --- | --- | --- | --- | --- |
| | | PACKAGED UNITS | | SPLIT SYSTEMS | |
| | | MINIMUM | MAXIMUM | MINIMUM | MAXIMUM |
| AIR CONDITIONER | SEER | 5.60 | 10.20 | 5.85 | 15.00 |
| HEAT PUMP | SEER | 6.50 | 10.50 | 7.25 | 13.05 |
| HEAT PUMP | HSPF | 5.05 | 7.80 | 5.30 | 8.90 |

Performance data for central air conditioners are shown in Section AC of the Directory. The cooling capacity and the SEER are listed for each single-package air conditioner and for each split-system combination of condensing (outdoor) unit and indoor coil.

Performance data for heat pumps are shown in Section HP of the Directory. Capacities and efficiencies for cooling and heating are listed for each single-package heat pump and for each split-system combination of outdoor and indoor coil. Cooling and heating efficiencies are expressed in SEER and HSPF, respectively.

The Directory lists the average national annual operating cost for each air-conditioner in Section AC. It also lists the average national annual operating cost for cooling and for heating in <u>Region IV</u> (see map, fig. 2) for each heat pump in Section HP.

Estimates of operating costs may be higher or lower than your average operating costs. They are affected by many factors that can vary widely. For example, since no two heating or cooling seasons are identical, operating costs will vary from year to year. Operating costs are also affected by the temperature that is to be maintained - the thermostat setting - with higher settings costing more in winter and lower settings costing more in summer. Other factors that affect system operation include the number of occupants, location within a region, activities that generate or release heat with the structure, and living habits such as the opening of windows, etc. Nevertheless, the estimates will be helpful in determining approximately how much a system will cost to operate and to compare the performance of different systems.

# ESTIMATING ANNUAL OPERATING COSTS

The average national annual operating costs listed in the Directory are based on U.S. Government standard tests and national averages of 1000 cooling load hours, 2080 heat load hours and national average electric rates of 8.30 cents per kWh. The step-by-step procedure calls for your local heating and cooling load hours and electric rates.

## STEP-BY-STEP PROCEDURE

The following is a step-by-step procedure for estimating your annual operating costs for cooling with a central air conditioner or heat pump and for heating when using a heat pump. Enter the required information in the accompanying worksheet. Letters and numbers in the procedure correspond to those on the worksheet. It will be helpful to look at the sample worksheet while you read the procedure.

A. The Building (home). Enter the identified information on the worksheet line indicated.

    1. Location -- the city and state in which the building is located.

    2. Building Heat Gain -- the calculated rate of heat flow from the environment, including heat gain from the sun, into the building at summer design conditions (sometimes called cooling load), expressed in Btu per hour (Btuh). Building heat gain should be calculated by an engineer, contractor or dealer.

    3. Building Heat Loss -- the calculated rate of heat flow from the building at winter design conditions, expressed in Btu per hour (Btuh). Building heat loss should be calculated by an engineer, contractor or dealer.

    4. Local power costs, dollars per kilowatt hour ($/kWh).

        a) Summer rate
        b) Winter rate

    Local power costs are available from your local power company or from your electric bill.

B. System Data. Locate the information requested below in Section AC of this Directory for air conditioners and Section HP for heat pumps.

    1. Equipment manufacturer's name.

    2. Model number, include both condensing unit and coil for split-systems air conditioners, or outdoor unit and indoor unit for split-system heat pumps.

    3. Unit cooling capacity, Btuh = Directory Cooling Capacity in MBtuh X 1000.

    4. Unit heating capacity at 47F for heat pump, Btuh = Directory Heating Capacity in MBtuh x 1000.

    5. Unit Average national annual operating costs, dollars.

        a) Cooling
        b) Heating

C. Operating Hours and Cost Factors. Locate the required information in the figures and tables of this Guide section of the Directory.

    1. Operating hours for your geographic location for cooling and heating.

        a) Summer cooling load hours from map, Fig. 1;
        b) Winter heating load hours from map, Fig. 2.

    2. Climatic Region, heating, from map Fig. 2 (Roman Numerals).

    3. Select Cooling Cost Factor from Table 1 as follows:

### STEP-BY-STEP PROCEDURE-Continued

      a) In the unit Capacity Column, locate the row with the capacity nearest your air conditioner or heat pump cooling capacity (line B.3 of worksheet). In the same row, select the cooling cost factor in the column with the heat gain nearest your building heat gain (line A.2 of worksheet). Enter the cooling cost factor on line C.3.a of worksheet.

    4. Select Heating Cost Factor from Table 2 as follows:

      a) Identify range of comparability, first column, that includes heat pump heating capacity (line b.4 of worksheet). Within that range, identify climatic region (line C.2 of worksheet). Select Heating Cost Factor in column headed by the value nearest your building heat loss (line A.3 of worksheet). Enter on Line C.4.a of worksheet.

D. Calculate estimated operating cost for cooling using the following equation:

Cooling Cost = ((A.4.a) (B.5.a)(C.1.a)(C.3.a)) / ((1000)(0.0830))

    Where:    A.4.a = line A.4.a = local power cost, summer, $/kWh

                  B.5.a = line B.5.a = unit average national annual cost cooling, $

                  C.1.a = line C.1.a = Summer cooling load hours.

                  C.3.a = line C.3.a = Cooling cost factor.

                  1000 = average national cooling cost factor.

                  0.0830 = national average power cost, $/kWh.

E. Calculate estimated operating cost for heating using the following equation:

Heating Cost = ((A.4.b) (B.5.b)(C.1.b)(C.4.a)) / ((2080)(0.0830))

    Where:    A.4.b = line A.4.b = local power cost, winter, $/kWh

                  B.5.b = line B.5.b = unit average national annual cost heating, $

                  C.1.b = line C.1.b = Winter heating load hours.

                  C.4.a = line C.4.a = Heating cost factor.

                  2080 = average national heating load hours, climate Region IV.

                  0.0830 = national average power cost, $/kWh.

F. Calculate estimated total operating cost for cooling and heating.

    Total Operating Cost = Cooling Cost + Heating Cost.

COMPLETED SAMPLE WORKSHEET
for
ESTIMATING ANNUAL OPERATING COST
OF AN
AIR CONDITIONER OR HEAT PUMP

A. **The Building**

1. Location  Fairfax Virginia
2. Building Heat Gain  33000  Btuh
3. Building Heat Loss  50000  Btuh
4. Local Power Rates

    a) Summer  0.0759  $/kWh    b) Winter  0.0716  $/kWh

B. **System Data from Directory**

1. Equipment Manufacturer  ABC Company
2. Model Number  XYZ-C8143
3. Unit Cooling Capacity  36.0 x 1000 = 36000
    (Directory Cooling Capacity, MBtuh) (1000) = Btuh
4. Unit Heating Capacity at 47 F  38.0 x 1000 = 38000
    (Directory Heating Capacity, MBtuh)(1000) = Btuh
5. Unit Average National Annual Operating Cost

    a) Cooling $ 345    b) Heating $ 728

C. **Operating Hours for Cost Factors**

1. Operating Hours for Cooling and Heating

    a) Summer Cooling Load Hours from Fig. 1  900

    b) Winter Heating Load Hours from Fig. 2  2100

2. Climatic Region, Heating, from Fig. 2  IV  (Roman Numeral)

3. Select Cooling Cost Factor from Table 1

    a) Cooling Cost Factor  0.93

4. Select Hating Cost Factor from Table 2

    a) Heating Cost Factor  1.266

D. **Calculate Estimated Cooling Cost**

   Cooling Cost = (( A.4.a ) ( B.5.a ) ( C.1.a ) ( C.3.a )) / (( 1,000 ) ( 0.0830 ))
   = ((0.0759) ( 345 ) ( 900 ) ( 0.93 )) / (( 1,000 ) ( 0.0830 ))
   = $ 264

E. **Calculate Estimated Heating Cost**

   Heating Cost = (( A.4.b ) ( B.5.b ) ( C.1.b ) ( C.4.a )) / (( 2,080 ) ( 0.0830 ))
   = ((0.0716) ( 728 ) ( 2100 ) ( 1.266 )) / (( 2,080 ) ( 0.0830 ))
   = $ 802

F. **Calculate Estimated Operating Cost for Heating and Cooling**

                          Line D      Line E
   Total Operating Cost = Cooling Cost + Heating Cost
   = ( 264 ) + ( 802 ) = $ 1066

OPERATING AT PEAK EFFICIENCY

## WORKSHEET for ESTIMATING ANNUAL OPERATING COST OF AN AIR CONDITIONER OR HEAT PUMP

**A. The Building**

1. Location _____

2. Building Heat Gain _____ Btuh

3. Building Heat Loss _____ Btuh

4. Local Power Rates

   a) Summer _____ $/kWh   b) Winter _____ $/kWh

**B. System Data from Directory**

1. Equipment Manufacturer _____

2. Model Number _____

3. Unit Cooling Capacity $\overline{\text{(Directory Cooling Capacity, MBtuh) (1000)}}$ = Btuh

4. Unit Heating Capacity at 47 F $\overline{\text{(Directory Heating Capacity, MBtuh)(1000)}}$ = Btuh

5. Unit Average National Annual Operating Cost

   a) Cooling $ _____   b) Heating $ _____

**C. Operating Hours for Cost Factors**

1. Operating Hours for Cooling and Heating

   a) Summer Cooling Load Hours from Fig. 1 _____

   b) Winter Heating Load Hours from Fig. 2 _____

2. Climatic Region, Heating, from Fig. 2 _____ (Roman Numeral)

3. Select Cooling Cost Factor from Table 1

   a) Cooling Cost Factor _____

4. Select Hating Cost Factor from Table 2

   a) Heating Cost Factor _____

**D. Calculate Estimated Cooling Cost**

Cooling Cost = (( A.4.a ) ( B.5.a ) ( C.1.a ) ( C.3.a )) / (( 1,000 ) ( 0.0830 ))
            = ((    ) (    ) (    ) (    )) / (( 1,000 ) ( 0.0830 ))
            = $ _____

**E. Calculate Estimated Heating Cost**

Heating Cost = (( A.4.b ) ( B.5.b ) ( C.1.b ) ( C.4.a )) / (( 2,080 ) ( 0.0830 ))
             = ((    ) (    ) (    ) (    )) / (( 2,080 ) ( 0.0830 ))
             = $ _____

**F. Calculate Estimated Operating Cost for Heating and Cooling**

$$\text{Total Operating Cost} = \underset{\text{Line D}}{\text{Cooling Cost}} + \underset{\text{Line E}}{\text{Heating Cost}}$$

= ( _____ ) + ( _____ ) = $ _____

## TABLE 2     HEATING COST FACTOR

| RANGE OF COMPARABILITY BTU/HR | | 5 | 10 | 15 | 20 | 25 | 30 | 35 | 40 | 50 | 60 | 70 | 80 | 90 | 100 | 110 | 130 |
|---|---|---|---|---|---|---|---|---|---|---|---|---|---|---|---|---|---|
| | | | | | | BUILDING HEAT LOSS, MBTU/HR (MBTU/HR times 1000 = BTU/HR) | | | | | | | | | | | |
| UP TO 12,499 | REGION I | 0.417 | 0.791 | 1.208 | | | | | | | | | | | | | |
| | REGION II | 0.444 | 0.836 | 1.266 | 1.795 | | | | | | | | | | | | |
| | REGION III | 0.474 | 0.893 | 1.330 | 1.856 | 2.482 | | | | | | | | | | | |
| | REGION IV | | 1.000 | 1.501 | 2.086 | 2.768 | 3.511 | | | | | | | | | | |
| | REGION V | | 1.166 | 1.747 | 2.389 | 3.098 | 3.862 | | | | | | | | | | |
| | REGION VI | 0.432 | 0.814 | 1.203 | 1.679 | | | | | | | | | | | | |
| 12,500-17,499 | REGION I | 0.293 | 0.558 | 0.820 | 1.113 | | | | | | | | | | | | |
| | REGION II | 0.311 | 0.590 | 0.859 | 1.158 | 1.506 | | | | | | | | | | | |
| | REGION III | | 0.624 | 0.907 | 1.210 | 1.557 | 1.953 | | | | | | | | | | |
| | REGION IV | | 0.682 | 1.000 | 1.343 | 1.729 | 2.163 | 2.636 | 3.137 | | | | | | | | |
| | REGION V | | 0.780 | 1.152 | 1.545 | 1.971 | 2.431 | 2.924 | 3.438 | | | | | | | | |
| | REGION VI | 0.303 | 0.577 | 0.836 | 1.111 | 1.428 | 1.785 | 2.172 | | | | | | | | | |
| 17,500-22,499 | REGION I | | 0.425 | 0.617 | 0.817 | 1.039 | | | | | | | | | | | |
| | REGION II | | 0.450 | 0.652 | 0.856 | 1.086 | 1.350 | | | | | | | | | | |
| | REGION III | | | 0.690 | 0.905 | 1.136 | 1.398 | 1.693 | 2.023 | | | | | | | | |
| | REGION IV | | | 0.755 | 1.000 | 1.266 | 1.561 | 1.888 | 2.249 | 3.037 | | | | | | | |
| | REGION V | | | 0.869 | 1.159 | 1.467 | 1.797 | 2.151 | 2.529 | 3.342 | 4.205 | | | | | | |
| | REGION VI | | 0.440 | 0.637 | 0.831 | 1.038 | 1.274 | 1.543 | 1.846 | | | | | | | | |
| 22,500-27,499 | REGION I | | 0.347 | 0.505 | 0.661 | 0.825 | 1.006 | | | | | | | | | | |
| | REGION II | | | 0.534 | 0.695 | 0.862 | 1.048 | 1.256 | 1.486 | | | | | | | | |
| | REGION III | | | 0.564 | 0.736 | 0.909 | 1.095 | 1.300 | 1.527 | 2.046 | | | | | | | |
| | REGION IV | | | | 0.803 | 1.000 | 1.211 | 1.441 | 1.690 | 2.257 | 2.883 | | | | | | |
| | REGION V | | | | 0.921 | 1.152 | 1.394 | 1.650 | 1.923 | 2.514 | 3.157 | 3.834 | 4.532 | | | | |
| | REGION VI | | | 0.523 | 0.680 | 0.837 | 1.005 | 1.193 | 1.400 | 1.879 | 2.374 | | | | | | |
| 27,500-32,499 | REGION I | | | 0.435 | 0.567 | 0.701 | 0.842 | | | | | | | | | | |
| | REGION II | | | 0.461 | 0.599 | 0.735 | 0.878 | 1.036 | 1.207 | | | | | | | | |
| | REGION III | | | 0.484 | 0.631 | 0.775 | 0.922 | 1.078 | 1.246 | 1.635 | | | | | | | |
| | REGION IV | | | | 0.677 | 0.836 | 1.000 | 1.175 | 1.363 | 1.788 | 2.276 | | | | | | |
| | REGION V | | | | 0.761 | 0.946 | 1.138 | 1.338 | 1.549 | 2.007 | 2.512 | 3.056 | 3.623 | | | | |
| | REGION VI | | | 0.451 | 0.587 | 0.719 | 0.853 | 0.997 | 1.156 | 1.512 | 1.919 | | | | | | |
| 32,500-37,499 | REGION I | | | 0.371 | 0.485 | 0.595 | 0.708 | 0.827 | | | | | | | | | |
| | REGION II | | | | 0.515 | 0.630 | 0.746 | 0.868 | 0.998 | 1.291 | | | | | | | |
| | REGION III | | | | 0.544 | 0.667 | 0.790 | 0.915 | 1.045 | 1.335 | 1.670 | | | | | | |
| | REGION IV | | | | | 0.723 | 0.859 | 1.000 | 1.147 | 1.470 | 1.835 | 2.244 | 2.675 | 3.133 | | | |
| | REGION V | | | | | 0.822 | 0.981 | 1.145 | 1.314 | 1.672 | 2.062 | 2.485 | 2.929 | 3.394 | | | |
| | REGION VI | | | | 0.503 | 0.617 | 0.728 | 0.841 | 0.962 | 1.230 | 1.532 | | | | | | |
| 37,500-44,999 | REGION I | | | | 0.328 | 0.429 | 0.528 | 0.625 | 0.725 | 0.831 | | | | | | | |
| | REGION II | | | | | 0.560 | 0.661 | 0.763 | 0.870 | 1.107 | 1.380 | | | | | | |
| | REGION III | | | | | 0.699 | 0.807 | 0.917 | 1.153 | 1.421 | | | | | | | |
| | REGION IV | | | | | | 0.876 | 1.000 | 1.266 | 1.563 | 1.898 | 2.269 | 2.662 | 3.077 | | | 3.784 |
| | REGION V | | | | | | 0.998 | 1.144 | 1.450 | 1.779 | 2.136 | 2.521 | 2.924 | 3.349 | 3.784 | | |
| | REGION VI | | | | | 0.548 | 0.647 | 0.746 | 0.845 | 1.061 | 1.310 | 1.590 | 1.902 | | | | |
| 45,000-49,999 | REGION I | | | | 0.349 | 0.429 | 0.508 | 0.587 | 0.669 | | | | | | | | |
| | REGION II | | | | | 0.456 | 0.538 | 0.620 | 0.703 | 0.881 | 1.086 | | | | | | |
| | REGION III | | | | | 0.568 | 0.654 | 0.741 | 0.921 | 1.122 | 1.348 | 1.600 | | | | | |
| | REGION IV | | | | | | 0.799 | 1.000 | 1.222 | 1.470 | 1.745 | 2.045 | 2.360 | 2.691 | | | |
| | REGION V | | | | | | 0.907 | 1.143 | 1.392 | 1.661 | 1.952 | 2.261 | 2.586 | 2.925 | 3.627 | | |
| | REGION VI | | | | | 0.446 | 0.527 | 0.607 | 0.687 | 0.853 | 1.038 | 1.248 | 1.477 | 1.726 | 1.978 | 2.232 | 2.808 |
| 50,000-54,999 | REGION I | | | | 0.353 | 0.435 | 0.515 | 0.594 | 0.673 | 0.840 | | | | | | | |
| | REGION II | | | | | 0.539 | 0.621 | 0.703 | 0.872 | 1.059 | 1.267 | | | | | | |
| | REGION III | | | | | 0.652 | 0.739 | 0.915 | 1.102 | 1.307 | 1.543 | 1.792 | | | | | |
| | REGION IV | | | | | | | 1.000 | 1.215 | 1.449 | 1.709 | 1.984 | 2.292 | 2.607 | | | |
| | REGION V | | | | | | | 1.156 | 1.408 | 1.671 | 1.957 | 2.254 | 2.573 | 2.900 | 3.597 | | |
| | REGION VI | | | | | 0.530 | 0.610 | 0.690 | 0.849 | 1.021 | 1.209 | 1.419 | 1.646 | 1.903 | 2.162 | 2.692 | |
| 55,000-59,999 | REGION I | | | | | 0.366 | 0.433 | 0.499 | 0.565 | 0.699 | | | | | | | |
| | REGION II | | | | | | 0.526 | 0.594 | 0.731 | 0.877 | 1.038 | | | | | | |
| | REGION III | | | | | | 0.626 | 0.770 | 0.917 | 1.076 | 1.254 | 1.447 | | | | | |
| | REGION IV | | | | | | | 0.834 | 1.000 | 1.179 | 1.374 | 1.585 | 1.817 | 2.060 | | | |
| | REGION V | | | | | | | 0.950 | 1.145 | 1.348 | 1.565 | 1.794 | 2.037 | 2.291 | 2.833 | | |
| | REGION VI | | | | | | 0.516 | 0.583 | 0.715 | 0.851 | 0.999 | 1.161 | 1.344 | 1.542 | | | |
| 60,000-64,999 | REGION I | | | | | | 0.370 | 0.438 | 0.505 | 0.572 | 0.706 | 0.849 | | | | | |
| | REGION II | | | | | | | 0.533 | 0.602 | 0.739 | 0.884 | 1.042 | 1.219 | | | | |
| | REGION III | | | | | | | 0.777 | 0.926 | 1.082 | 1.254 | 1.447 | 1.656 | | | | |
| | REGION IV | | | | | | | | 1.000 | 1.177 | 1.371 | 1.581 | 1.808 | 2.055 | 2.590 | | |
| | REGION V | | | | | | | | | 1.140 | 1.345 | 1.563 | 1.792 | 2.037 | 2.293 | 2.841 | |
| | REGION VI | | | | | | 0.523 | 0.591 | 0.725 | 0.859 | 1.005 | 1.164 | 1.341 | 1.534 | | | |

## TABLE 1    COOLING COST FACTOR

| RANGE OF COMPARABILITY BTU/HR | UNIT CAPACITY BTU/HR | BUILDING HEAT GAIN BTU/HR | | |
|---|---|---|---|---|
| 7,500-12,499 | | 7,500 | 10,000 | 12,500 |
| | 7,500 | 1.00 | 1.33 | 1.67 |
| | 10,000 | 0.75 | 1.00 | 1.25 |
| | 12,499 | 0.60 | 0.80 | 1.00 |
| 12,500-17,499 | | 12,500 | 15,000 | 17,500 |
| | 12,500 | 1.00 | 1.20 | 1.40 |
| | 15,000 | 0.83 | 1.00 | 1.17 |
| | 17,499 | 0.71 | 0.86 | 1.00 |
| 17,500-22,499 | | 17,500 | 20,000 | 22,500 |
| | 17,500 | 1.00 | 1.14 | 1.29 |
| | 20,000 | 0.88 | 1.00 | 1.13 |
| | 22,499 | 0.78 | 0.89 | 1.00 |
| 22,500-27,499 | | 22,500 | 25,000 | 27,500 |
| | 22,500 | 1.00 | 1.11 | 1.22 |
| | 25,000 | 0.90 | 1.00 | 1.10 |
| | 27,499 | 0.82 | 0.91 | 1.00 |
| 27,500-32,499 | | 27,500 | 30,000 | 32,500 |
| | 27,500 | 1.00 | 1.09 | 1.18 |
| | 30,000 | 0.92 | 1.00 | 1.08 |
| | 32,499 | 0.85 | 0.92 | 1.00 |
| 32,500-37,499 | | 32,500 | 35,000 | 37,500 |
| | 32,500 | 1.00 | 1.08 | 1.15 |
| | 35,000 | 0.93 | 1.00 | 1.07 |
| | 37,500 | 0.87 | 0.93 | 1.00 |
| 37,500-44,999 | | 37,500 | 41,250 | 45,000 |
| | 37,500 | 1.00 | 1.10 | 1.20 |
| | 41,250 | 0.91 | 1.00 | 1.09 |
| | 45,000 | 0.83 | 0.92 | 1.00 |
| 45,000-49,499 | | 45,000 | 47,500 | 50,000 |
| | 45,000 | 1.00 | 1.06 | 1.11 |
| | 47,500 | 0.95 | 1.00 | 1.05 |
| | 49,999 | 0.90 | 0.95 | 1.00 |
| 50,000-54,999 | | 50,000 | 52,500 | 55,000 |
| | 50,000 | 1.00 | 1.05 | 1.10 |
| | 52,500 | 0.95 | 1.00 | 1.05 |
| | 54,999 | 0.91 | 0.95 | 1.00 |
| 55,000-59,999 | | 55,000 | 57,500 | 60,000 |
| | 55,000 | 1.00 | 1.05 | 1.09 |
| | 57,500 | 0.96 | 1.00 | 1.04 |
| | 59,999 | 0.92 | 0.96 | 1.00 |
| 60,000-64,999 | | 60,000 | 62,500 | 65,000 |
| | 60,000 | 1.00 | 1.04 | 1.08 |
| | 62,500 | 0.96 | 1.00 | 1.04 |
| | 65,000 | 0.92 | 0.96 | 1.00 |

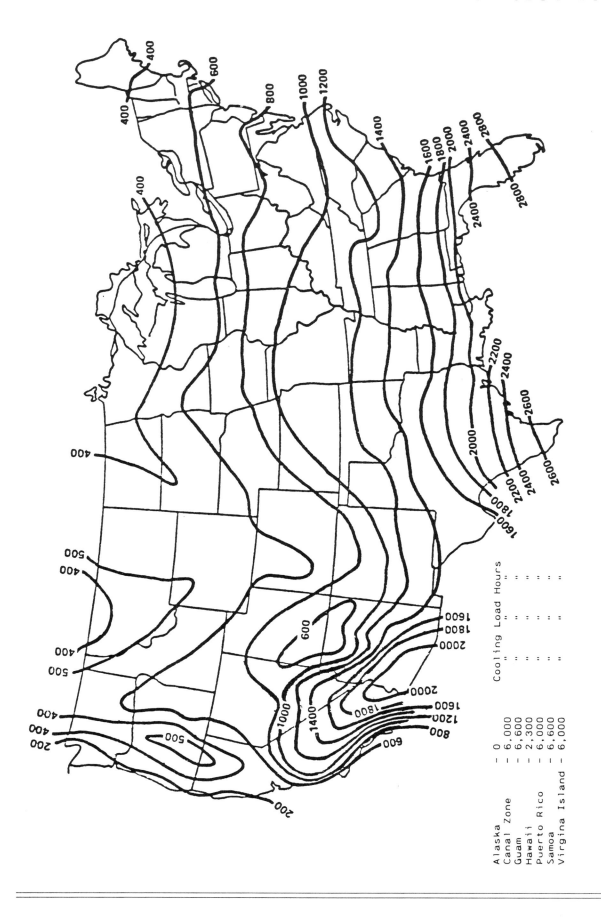

FIG 1. SUMMER COOLING LOAD HOURS

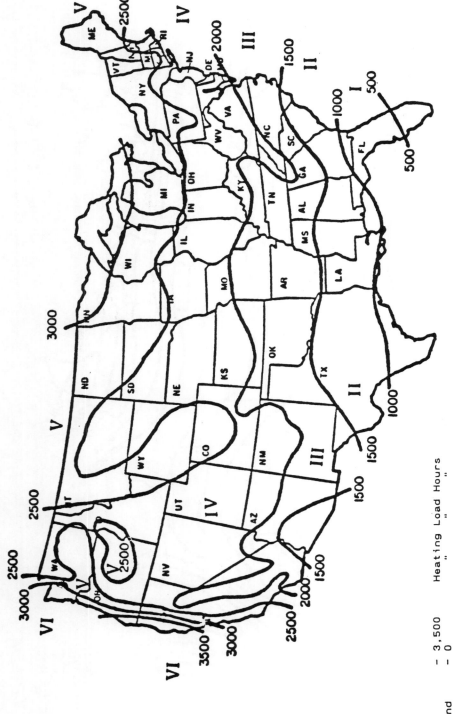

FIG 2. WINTER HEATING LOAD HOURS

# Chapter 11

# Estimating Annual Heating Requirements

The following pages are reprinted with permission from the Gas Appliance Manufacturers Association (GAMA), from the *Consumer's Directory of Certified Efficiency Ratings for Residential Heating and Water Heating Equipment*, April, 1992.

**Procedure for (a) Estimating the Annual Heating Requirements and (b) Comparing the Costs of Operation of Different Models**

To estimate the amount of energy a home will use to provide comfort heating for a year, it is necessary to use both a) information specific to the particular home's need for heat and b) the size and efficiency information on models listed in this Directory.

Outlined below is a method (based on the Department of Energy efficiency test procedure for furnaces and boilers) for first estimating the annual amount of energy[1] used for heating a specific installation and then making a comparison of the estimated annual operating costs of models of essentially the same size but different efficiencies. Because of the number of variables involved in the procedure, the method shown here is only for the purpose of 1) estimating the amount of energy that will be consumed in one year for the specific installation and 2) comparing models of various Annual Fuel Utilization Efficiencies (AFUE) to assist in making a purchasing decision. The method outlined below can be used for comparing models using the same fuel (gas to gas) or different fuels (gas to oil). When selecting models for comparison, make sure units are of the same configuration needed for a particular installation, i.e., upflow models, etc. A "Worksheet" is provided at the end of the section for ease in compiling and calculating the necessary information. The steps required to complete the "Worksheet"

Step 1 - Determine the Heating Load Hours (HLH) for your area. To find an approximate number of Heating Load Hours for your specific area, use the map illustrated in Figure 1.

> Heating Load Hours, HLH _____ hours

(1)
The procedure outlined above is based upon the Department of Energy test procedure for estimating the annual operating cost of gas or oil-fired furnaces or boilers. It includes the cost of annual fuel usage (natural gas, propane gas, or No. 2 heating oil), in addition to the annual electrical cost to operate furnace circulating blowers, pumps, and power burners.

# APRIL 1992                                                                 PAGE 5

Step 2 - Estimate the Design Heating Requirement (DHR) for the specific installation. Because the Design Heating Requirement is dependent upon a number of variables, such as the size of the house, building materials, insulation, architectural features, specific climatic conditions, etc., this determination must usually be performed by a knowledgeable heating contractor or other qualified source.

> Design Heating Requirement, DHR _____ Btu/hr

Step 3 - Refer to the Directory to identify the size of model needed to satisfy the Design Heating Requirement determined in Step 2. Models listed in the Directory of essentially the same heating capacity can now be evaluated for estimated annual operating costs. Models of the same fuel type (gas to gas) as well as different fuel types (gas to oil) can be compared using this procedure.

> Input _____ Btu/hr
> Heating Capacity _____ Btu/hr
> AFUE _____ %
> Average Annual Fuel Consumption, Ef _____ MMBtu
> Average Annual Electrical Consumption, Eae _____ Kw-hr

Step 4 - Determine the rated Design Heating Requirement, RDHR, for the selected model. The RDHR for an appliance is the DHR value which was used to calculate the Ef and Eae ratings included in this Directory. Using the Heating Capacity from Step 3, read the RDHR from Table 1.

> Rated Design Heating Requirement, RDHR _____ Btu/hr

Step 5 - Calculate the adjustment factor, AF, which is required to correct the energy usage figures from the Directory for your specific installation.

$$AF = \frac{HLH \times DHR}{2080 \times RDHR}$$

where,
   HLH = Heating Load Hours from Step 1.
   DHR = Design Heating Requirements from Step 2.
   2080 = Average Annual Heating Load hours.
   RDHR = Rated Design Heating Requirement from Step 4.

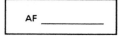

> AF _____

Step 6 - Calculate the Estimated Annual Fuel Usage, EAFU, for your specific installation.

   EAFU = AF x Ef

where,
   AF = Adjustment Factor from Step 5.
   Ef = Average Annual Fuel Consumption from Step 3.

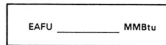

> EAFU _____ MMBtu

Step 7 - Calculate the Estimated Annual Electrical Usage, EAEU, for your specific installation.

   EAEU = AF x Eae

where,
   AF = Adjustment Factor from Step 5.
   Eae = Average Annual Electrical Consumption from Step 3.

> EAEU _____ kw-hrs

Step 8 - Calculate the Estimated Annual Operating Cost, EAOC, for your specific installation and selected model.

$$EAOC = \frac{(EAFU \times 1{,}000{,}000 \times Fuel\ Cost)}{BTU\ Content} + (EAEU \times Electrical\ Cost)$$

where,
- EAFU = Estimated Annual Fuel Usage in MMBtu from Step 6.
- 1,000,000 = Conversion factor for MMBtu to BTU.
- Fuel Cost = Cost of fuel for your area in:
    - $ per therm for Natural Gas.
    - $ per gallon for Propane Gas.
    - $ per gallon for Heating Oil.
- BTU Content =
    - 100,000 Btu per therm for Natural Gas.
    - 91,000 Btu per gallon for Propane Gas.
    - 138,700 Btu per gallon for Heating Oil.
- EAEU = Estimated Annual Electrical Usage in kw-hr from Step 7.
- Electrical Cost = Cost of electricity in your area in $ per kw-hr.

EAOC, $ _____ per Year

This procedure can be repeated for other models of essentially the same size with different efficiencies to compare their Estimated Annual Operating Costs. Models with higher efficiency (AFUE) ratings will consume less fuel and cost less to operate, but generally have a higher purchase price and may have a higher installation cost. Therefore, there is a period of time, referred to as "Payback Period", before the savings that results from the lower operating costs of a more efficient model makes up the difference in price of that furnace or boiler as compared to a less efficient model.

EXAMPLE: Assume you are intending to buy a new gas furnace and have calculated the Estimated Annual Operating Cost (EAOC) using the procedure outlined above for two models of essentially the same size to meet your heating requirements but with different efficiency (AFUE) ratings.

|  | Price of Furnace | AFUE | Estimated Annual Operating Cost |
|---|---|---|---|
| Model A | $600 Installed | 78% | $425.00 |
| Model B | $775 Installed | 91% | $365.00 |

The additional cost of more efficient model (Model B) is: $775 - 600 = $175 higher installation cost.

The estimated Annual Savings in Operating Costs for Model B, as compared to Model A, is:

$425 - 365 = $60 lower operating cost per year.

The Payback Period is the ratio of the higher installed cost to the lower annual operating cost:

PAYBACK PERIOD = $175 / $60 = 2.9 years

### TABLE 1

| Unit Heating Capacity, BTUH | RDHR, BTUH |
|---|---|
| 5,000 - 10,000 | 5,000 |
| 11,000 - 16,000 | 10,000 |
| 17,000 - 25,000 | 15,000 |
| 26,000 - 42,000 | 20,000 |
| 43,000 - 59,000 | 30,000 |
| 60,000 - 76,000 | 40,000 |
| 77,000 - 93,000 | 50,000 |
| 94,000 - 110,000 | 60,000 |
| 111,000 - 127,000 | 70,000 |
| 128,000 - 144,000 | 80,000 |
| 145,000 - 161,000 | 90,000 |
| 162,000 - 178,000 | 100,000 |
| 179,000 - 195,000 | 110,000 |
| 196,000 and Over | 130,000 |

## Heating Load Hours (HLH) for the United States

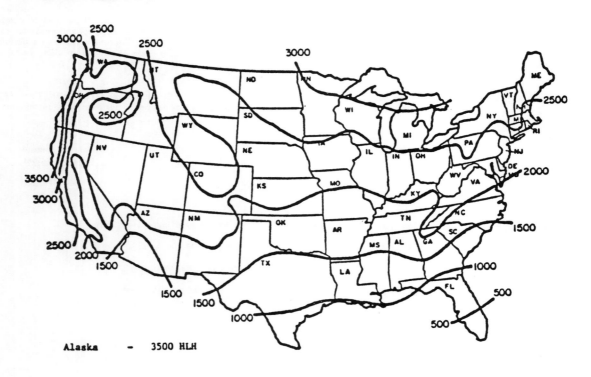

This map is reasonably accurate for most parts of the United States but is necessarily highly generalized and consequently not too accurate in mountainous regions, particularly in the Rockies.

FIGURE 1

ESTIMATING ANNUAL HEATING REQUIREMENTS

WORKSHEET

General instructions: After determining the size of the unit needed, refer to the listing of models of essentially the same size (heating capacity) and select models for comparison. Fill in the blocks with the listing information and perform the calculations required to determine the EAOC parameter - Estimated Annual Operating Costs.

|  | MODEL 1 | MODEL 2 |
|---|---|---|
| Brand: <br> Model: | | |
| STEP 1. Heating Load Hours (Figure 1) | | |
| STEP 2. Design Heating Requirements, DHR, in BTU/hr | | |
| STEP 3. Record the following information from the Directory: <br> Input, BTU/hr <br> Heating Capacity, BTU/hr <br> Afue, % <br> Ef, MMBtu <br> Eae, kw-hr | | |
| STEP 4. Rated Design Heating Requirement, RDHR, from Table 1. | | |
| STEP 5. Calculate the value AF: $$AF = \frac{HLH \times DHR}{2080 \times RDHR}$$ | | |
| STEP 6. Calculate Estimated Annual Fuel Usage, EAFU, in MMbtu's: $$EAFU = AF \times Ef$$ | | |
| STEP 7. Calculate Estimated Annual Electrical Usage, EAEU, in kw-hrs: $$EAEU = AF \times Eae$$ | | |
| STEP 8. Calculate Estimated Annual Operating Cost, EAOC, in $: $$EAOC = \frac{(EAFU \times 1{,}000{,}000 \times Cost_{Fuel})}{BTU\ Content} + (EAEU \times Cost_{Elec})$$ <br> Natural Gas: $0.6054/therm @ 100,000 Btu/therm <br> Propane Gas: $0.89/gallon @ 91,000 Btu/gal <br> Heating Oil: $1.29/gallon @ 138,700 Btu/gal <br> National Average Electric Cost = $0.0824/Kw-hr <br><br> Costs Based on FTC 1991 National Averages <br> Use Local Energy Costs When Available | | |

## DEFINITIONS

The following are some definitions of the terms used in this procedure:

1. Btu: British thermal units

2. Btu/hr or Btuh: Btu per hour

3. MBtu: Thousands of Btu; used to measure energy. For example, a few hundred Btu provide enough energy to make a pot of coffee, but some larger homes may require as much as 80,000 to 90,000 Btuh for heating on very cold days; 80,000 Btu = 80 MBtu.

4. MMBtu: Millions of Btu; used to measure energy; 8,000,000 Btu = 8 MMBtu

5. Input, Btuh: Represents the amount of fuel the model consumes in one hour. It defines the rate of energy supplied in a fuel to a furnace or boiler when operating under continuous burning (steady-state) conditions.

6. Steady-State Conditions: Conditions of continuous burner operation during which fuel consumption by the furnace or boiler is measured; somewhat like measuring an automobile's gasoline mileage under steady highway driving conditions.

7. Heating Capacity, MBtuh: Represents how much heat the model can produce in one hour operating under steady-state conditions; expressed in thousands of Btu per hour. For isolated combustion or outdoor units, the heating capacity is determined by multiplying the specified input by the steady-state efficiency, as tested, and then subtracting an additional therm to account for jacket loss, which would go into an unheated environment.

8. PE Watts: The electrical energy input rate supplied to the power burner (combustion air blower, fuel pump, damper motor) of a furnace or boiler operating under continuous burning (steady-state) conditions.

9. $E_{ae}$, Kw-hr/yr: The average annual auxiliary electrical energy consumption for a gas furnace or boiler in kilowatt-hours per year. It is a measure of the total electrical energy supplied to a furnace or boiler during a one year period.

10. $E_f$, MMBtu/yr: The average annual fuel energy consumption for a gas furnace or boiler; expressed in millions of Btu per year.

11. AFUE, %: AFUE stands for annual fuel utilization efficiency and is the efficiency rating of the model shown. Unlike steady-

state conditions, this rating is based on average usage, including on and off cycling, as set out in the standardized Department of Energy test procedures. The higher the AFUE rating, the more efficient the model will be.

Notes: a) Items 7 and 11 are certified values, while Items 8, 9, and 10 are provided for application use; b) Eae and Ef are based on national averages. For the purpose of comparison, use the calculation procedure in this section.

# Appendix A

# Air Conditioning Formulas (Non-Psychrometric)

1. Latent heat removed =

    Gallons of water per hour (gph) x 8830

2. Mixed air temperature (db) =

    $\dfrac{\text{Outdoor air}}{\text{Total air}}$ x Outdoor temperature + $\dfrac{\text{Return air}}{\text{Total air temperature}}$ x Return air temperature

3. Mixed air temperature (wb) =

    $\dfrac{\text{Outdoor air}}{\text{Total air}}$ x Outdoor heat content + $\dfrac{\text{Return air}}{\text{Total air}}$ x Heat content

4. Sensible heat (%) =

    $\dfrac{\text{Sensible heat}}{\text{Total heat}}$

5. Sensible heat removed =

    cfm x 1.08 x Change of db temperature

    where: 1.08 = specific heat of air constant (4.5 x 0.24)

6. Total heat removed =

    cfm x 4.5 x Change of enthalpy (Heat content)

    where: 4.5 = a constant (60 ÷ 13.33; 60 min = 1 hr, 13.33 = cu ft/lb dry air @ 70°F db)

7. 0.24 = Btu required to raise one pound of air 1°F

8. 1060 = Btu removed to condense one pound of water

9. 7000 = grains of water vapor per pound

10. 0.68 = Btu required to remove one grain of water from one cu ft of air:

    $$0.68 = 4.5 \times \frac{1060}{7000}$$

11. 8.33 lb = one gallon of water

# Appendix B

## Worksheets

# Electric Heating Unit Capacity Worksheet

Introduction: When a customer complains about insufficient heat from an electric heating unit, the service technician should, as a first step, determine if the heating elements are delivering the amount of heat they were designed to deliver. It could be that one or more of the heating elements is not operating properly. This is a fairly simple test and is easily performed.

Tools Needed: ammeter, voltmeter, accurate thermometer, and tool kit. An accurate wattmeter may be used in place of the ammeter and the voltmeter.

Procedures:

1. Turn off the electricity to any other equipment that is used in conjunction with the electric heating unit.

2. Set the thermostat to demand heat. Allow the heater to operate for approximately 10 minutes so the temperatures will stabilize.

3. Measure the voltage and amperage of each heating element and record:

|        | Volts | Amps |
|--------|-------|------|
| Motor: 1. | _____ | _____ |
| Heater: 1. | _____ | _____ |
| 2. | _____ | _____ |
| 3. | _____ | _____ |
| 4. | _____ | _____ |
| 5. | _____ | _____ |
| Sum: | _____ | _____ |

4. Determine the total wattage of the heat strips and record. Use the following formula:

$$W = V \times I$$

$$W = _____$$

5. Determine the Btu output of the heat strips and record. Use the following formula:

$$Btu = W \times 3.413$$

$$Btu = _____$$

Copyright © Business News Publishing Company

6. Determine the cfm of the blower and record. Use the following instructions:

   - Use the same thermometer, or two that measure exactly the same, to measure the return and supply air temperatures.

   - Do not measure the temperature in an area where the thermometer can sense the radiant heat from the heat strips (see Figure 2-2). True air temperature cannot be measured if the thermometer senses radiant heat.

   - Take the temperature measurements within 6 ft of the air handler. Measurements taken at the return and supply grilles that are at too great a distance from the unit are not usually accurate enough.

   - Use the average temperature when more than one duct is connected to the supply air plenum.

   - Be sure the air temperature has stabilized before taking the temperature measurements.

   - Take the temperature measurements downstream from any source of mixed air. Use the following formula:

   $$cfm = \frac{Btu}{1.08 \times \Delta T}$$

   cfm = _____

7. Is this what the manufacturer rates the equipment? _____

Copyright © Business News Publishing Company

# Electric Heating Unit Capacity Worksheet

Introduction: When a customer complains about insufficient heat from an electric heating unit, the serv. technician should, as a first step, determine if the heating elements are delivering the amount of heat they were designed to deliver. It could be that one or more of the heating elements is not operating properly. This is a fairly simple test and is easily performed.

Tools Needed: ammeter, voltmeter, accurate thermometer, and tool kit. An accurate wattmeter may be used in place of the ammeter and the voltmeter.

Procedures:

1. Turn off the electricity to any other equipment that is used in conjunction with the electric heating unit.

2. Set the thermostat to demand heat. Allow the heater to operate for approximately 10 minutes so the temperatures will stabilize.

3. Measure the voltage and amperage of each heating element and record:

    |        | Volts | Amps |
    |--------|-------|------|
    | Motor: 1. | _____ | _____ |
    | Heater: 1. | _____ | _____ |
    | 2. | _____ | _____ |
    | 3. | _____ | _____ |
    | 4. | _____ | _____ |
    | 5. | _____ | _____ |
    | Sum: | _____ | _____ |

4. Determine the total wattage of the heat strips and record. Use the following formula:

    $$W = V \times I$$

    W = _____

5. Determine the Btu output of the heat strips and record. Use the following formula:

    $$Btu = W \times 3.413$$

    Btu = _____

Copyright © Business News Publishing Company

6. Determine the cfm of the blower and record. Use the following instructions:

   - Use the same thermometer, or two that measure exactly the same, to measure the return and supply air temperatures.

   - Do not measure the temperature in an area where the thermometer can sense the radiant heat from the heat strips (see Figure 2-2). True air temperature cannot be measured if the thermometer senses radiant heat.

   - Take the temperature measurements within 6 ft of the air handler. Measurements taken at the return and supply grilles that are at too great a distance from the unit are not usually accurate enough.

   - Use the average temperature when more than one duct is connected to the supply air plenum.

   - Be sure the air temperature has stabilized before taking the temperature measurements.

   - Take the temperature measurements downstream from any source of mixed air. Use the following formula:

   $$cfm = \frac{Btu}{1.08 \times \Delta T}$$

   cfm = _____

7. Is this what the manufacturer rates the equipment? _____

# Electric Heating Unit Capacity Worksheet

Introduction: When a customer complains about insufficient heat from an electric heating unit, the service technician should, as a first step, determine if the heating elements are delivering the amount of heat they were designed to deliver. It could be that one or more of the heating elements is not operating properly. This is a fairly simple test and is easily performed.

Tools Needed: ammeter, voltmeter, accurate thermometer, and tool kit. An accurate wattmeter may be used in place of the ammeter and the voltmeter.

Procedures:

1. Turn off the electricity to any other equipment that is used in conjunction with the electric heating unit.

2. Set the thermostat to demand heat. Allow the heater to operate for approximately 10 minutes so the temperatures will stabilize.

3. Measure the voltage and amperage of each heating element and record:

    |         |    | Volts | Amps |
    |---------|----|-------|------|
    | Motor:  | 1. |       |      |
    | Heater: | 1. |       |      |
    |         | 2. |       |      |
    |         | 3. |       |      |
    |         | 4. |       |      |
    |         | 5. |       |      |
    | Sum:    |    |       |      |

4. Determine the total wattage of the heat strips and record. Use the following formula:

    $$W = V \times I$$

    W = _____

5. Determine the Btu output of the heat strips and record. Use the following formula:

    $$Btu = W \times 3.413$$

    Btu = _____

6. Determine the cfm of the blower and record. Use the following instructions:

    - Use the same thermometer, or two that measure exactly the same, to measure the return and supply air temperatures.

    - Do not measure the temperature in an area where the thermometer can sense the radiant heat from the heat strips (see Figure 2-2). True air temperature cannot be measured if the thermometer senses radiant heat.

    - Take the temperature measurements within 6 ft of the air handler. Measurements taken at the return and supply grilles that are at too great a distance from the unit are not usually accurate enough.

    - Use the average temperature when more than one duct is connected to the supply air plenum.

    - Be sure the air temperature has stabilized before taking the temperature measurements.

    - Take the temperature measurements downstream from any source of mixed air. Use the following formula:

    $$cfm = \frac{Btu}{1.08 \times \Delta T}$$

    cfm = _____

7. Is this what the manufacturer rates the equipment?_____

# Electric Heating Unit Capacity Worksheet

Introduction: When a customer complains about insufficient heat from an electric heating unit, the service technician should, as a first step, determine if the heating elements are delivering the amount of heat they were designed to deliver. It could be that one or more of the heating elements is not operating properly. This is a fairly simple test and is easily performed.

Tools Needed: ammeter, voltmeter, accurate thermometer, and tool kit. An accurate wattmeter may be used in place of the ammeter and the voltmeter.

Procedures:

1. Turn off the electricity to any other equipment that is used in conjunction with the electric heating unit.

2. Set the thermostat to demand heat. Allow the heater to operate for approximately 10 minutes so the temperatures will stabilize.

3. Measure the voltage and amperage of each heating element and record:

| | Volts | Amps |
|---|---|---|
| Motor: 1. | _____ | _____ |
| Heater: 1. | _____ | _____ |
| 2. | _____ | _____ |
| 3. | _____ | _____ |
| 4. | _____ | _____ |
| 5. | _____ | _____ |
| Sum: | _____ | _____ |

4. Determine the total wattage of the heat strips and record. Use the following formula:

$$W = V \times I$$

W = _____

5. Determine the Btu output of the heat strips and record. Use the following formula:

$$Btu = W \times 3.413$$

Btu = _____

6. Determine the cfm of the blower and record. Use the following instructions:

    - Use the same thermometer, or two that measure exactly the same, to measure the return and supply air temperatures.

    - Do not measure the temperature in an area where the thermometer can sense the radiant heat from the heat strips (see Figure 2-2). True air temperature cannot be measured if the thermometer senses radiant heat.

    - Take the temperature measurements within 6 ft of the air handler. Measurements taken at the return and supply grilles that are at too great a distance from the unit are not usually accurate enough.

    - Use the average temperature when more than one duct is connected to the supply air plenum.

    - Be sure the air temperature has stabilized before taking the temperature measurements.

    - Take the temperature measurements downstream from any source of mixed air. Use the following formula:

    $$cfm = \frac{Btu}{1.08 \times \Delta T}$$

    cfm = _____

7. Is this what the manufacturer rates the equipment?_____

# Electric Heating Unit Capacity Worksheet

Introduction: When a customer complains about insufficient heat from an electric heating unit, the service technician should, as a first step, determine if the heating elements are delivering the amount of heat they were designed to deliver. It could be that one or more of the heating elements is not operating properly. This is a fairly simple test and is easily performed.

Tools Needed: ammeter, voltmeter, accurate thermometer, and tool kit. An accurate wattmeter may be used in place of the ammeter and the voltmeter.

Procedures:

1. Turn off the electricity to any other equipment that is used in conjunction with the electric heating unit.

2. Set the thermostat to demand heat. Allow the heater to operate for approximately 10 minutes so the temperatures will stabilize.

3. Measure the voltage and amperage of each heating element and record:

|  |  | Volts | Amps |
|---|---|---|---|
| Motor: | 1. | _____ | _____ |
| Heater: | 1. | _____ | _____ |
|  | 2. | _____ | _____ |
|  | 3. | _____ | _____ |
|  | 4. | _____ | _____ |
|  | 5. | _____ | _____ |
| Sum: |  | _____ | _____ |

4. Determine the total wattage of the heat strips and record. Use the following formula:

$$W = V \times I$$

W = _____

5. Determine the Btu output of the heat strips and record. Use the following formula:

$$Btu = W \times 3.413$$

Btu = _____

6. Determine the cfm of the blower and record. Use the following instructions:

    - Use the same thermometer, or two that measure exactly the same, to measure the return and supply air temperatures.

    - Do not measure the temperature in an area where the thermometer can sense the radiant heat from the heat strips (see Figure 2-2). True air temperature cannot be measured if the thermometer senses radiant heat.

    - Take the temperature measurements within 6 ft of the air handler. Measurements taken at the return and supply grilles that are at too great a distance from the unit are not usually accurate enough.

    - Use the average temperature when more than one duct is connected to the supply air plenum.

    - Be sure the air temperature has stabilized before taking the temperature measurements.

    - Take the temperature measurements downstream from any source of mixed air. Use the following formula:

    $$\text{cfm} = \frac{\text{Btu}}{1.08 \times \Delta T}$$

    cfm = _____

7. Is this what the manufacturer rates the equipment?_____

# Electric Heating Unit Capacity Worksheet

Introduction: When a customer complains about insufficient heat from an electric heating unit, the service technician should, as a first step, determine if the heating elements are delivering the amount of heat they were designed to deliver. It could be that one or more of the heating elements is not operating properly. This is a fairly simple test and is easily performed.

Tools Needed: ammeter, voltmeter, accurate thermometer, and tool kit. An accurate wattmeter may be used in place of the ammeter and the voltmeter.

Procedures:

1. Turn off the electricity to any other equipment that is used in conjunction with the electric heating unit.

2. Set the thermostat to demand heat. Allow the heater to operate for approximately 10 minutes so the temperatures will stabilize.

3. Measure the voltage and amperage of each heating element and record:

|  |  | Volts | Amps |
|---|---|---|---|
| Motor: | 1. | _____ | _____ |
| Heater: | 1. | _____ | _____ |
|  | 2. | _____ | _____ |
|  | 3. | _____ | _____ |
|  | 4. | _____ | _____ |
|  | 5. | _____ | _____ |
| Sum: |  | _____ | _____ |

4. Determine the total wattage of the heat strips and record. Use the following formula:

$$W = V \times I$$

W = _____

5. Determine the Btu output of the heat strips and record. Use the following formula:

$$Btu = W \times 3.413$$

Btu = _____

Copyright © Business News Publishing Company

6. Determine the cfm of the blower and record. Use the following instructions:

    - Use the same thermometer, or two that measure exactly the same, to measure the return and supply air temperatures.

    - Do not measure the temperature in an area where the thermometer can sense the radiant heat from the heat strips (see Figure 2-2). True air temperature cannot be measured if the thermometer senses radiant heat.

    - Take the temperature measurements within 6 ft of the air handler. Measurements taken at the return and supply grilles that are at too great a distance from the unit are not usually accurate enough.

    - Use the average temperature when more than one duct is connected to the supply air plenum.

    - Be sure the air temperature has stabilized before taking the temperature measurements.

    - Take the temperature measurements downstream from any source of mixed air. Use the following formula:

    $$\text{cfm} = \frac{\text{Btu}}{1.08 \times \Delta T}$$

    cfm = _____

7. Is this what the manufacturer rates the equipment? _____

Copyright © Business News Publishing Company

# Electric Heating Unit Capacity Worksheet

Introduction: When a customer complains about insufficient heat from an electric heating unit, the service technician should, as a first step, determine if the heating elements are delivering the amount of heat they were designed to deliver. It could be that one or more of the heating elements is not operating properly. This is a fairly simple test and is easily performed.

Tools Needed: ammeter, voltmeter, accurate thermometer, and tool kit. An accurate wattmeter may be used in place of the ammeter and the voltmeter.

Procedures:

1. Turn off the electricity to any other equipment that is used in conjunction with the electric heating unit.

2. Set the thermostat to demand heat. Allow the heater to operate for approximately 10 minutes so the temperatures will stabilize.

3. Measure the voltage and amperage of each heating element and record:

|  | Volts | Amps |
|---|---|---|
| Motor: 1. | _____ | _____ |
| Heater: 1. | _____ | _____ |
| 2. | _____ | _____ |
| 3. | _____ | _____ |
| 4. | _____ | _____ |
| 5. | _____ | _____ |
| Sum: | _____ | _____ |

4. Determine the total wattage of the heat strips and record. Use the following formula:

$$W = V \times I$$

$W = _____$

5. Determine the Btu output of the heat strips and record. Use the following formula:

$$Btu = W \times 3.413$$

$Btu = _____$

6. Determine the cfm of the blower and record. Use the following instructions:

    - Use the same thermometer, or two that measure exactly the same, to measure the return and supply air temperatures.

    - Do not measure the temperature in an area where the thermometer can sense the radiant heat from the heat strips (see Figure 2-2). True air temperature cannot be measured if the thermometer senses radiant heat.

    - Take the temperature measurements within 6 ft of the air handler. Measurements taken at the return and supply grilles that are at too great a distance from the unit are not usually accurate enough.

    - Use the average temperature when more than one duct is connected to the supply air plenum.

    - Be sure the air temperature has stabilized before taking the temperature measurements.

    - Take the temperature measurements downstream from any source of mixed air. Use the following formula:

    $$\text{cfm} = \frac{\text{Btu}}{1.08 \times \Delta T}$$

    cfm = _____

7. Is this what the manufacturer rates the equipment?_____

# Electric Heating Unit Capacity Worksheet

Introduction: When a customer complains about insufficient heat from an electric heating unit, the service technician should, as a first step, determine if the heating elements are delivering the amount of heat they were designed to deliver. It could be that one or more of the heating elements is not operating properly. This is a fairly simple test and is easily performed.

Tools Needed: ammeter, voltmeter, accurate thermometer, and tool kit. An accurate wattmeter may be used in place of the ammeter and the voltmeter.

Procedures:

1. Turn off the electricity to any other equipment that is used in conjunction with the electric heating unit.

2. Set the thermostat to demand heat. Allow the heater to operate for approximately 10 minutes so the temperatures will stabilize.

3. Measure the voltage and amperage of each heating element and record:

|        |    | Volts | Amps |
|--------|----|-------|------|
| Motor: | 1. _____ | _____ |
| Heater: | 1. _____ | _____ |
|        | 2. _____ | _____ |
|        | 3. _____ | _____ |
|        | 4. _____ | _____ |
|        | 5. _____ | _____ |
| Sum: | _____ | _____ |

4. Determine the total wattage of the heat strips and record. Use the following formula:

$$W = V \times I$$

W = _____

5. Determine the Btu output of the heat strips and record. Use the following formula:

$$Btu = W \times 3.413$$

Btu = _____

6. Determine the cfm of the blower and record. Use the following instructions:

   - Use the same thermometer, or two that measure exactly the same, to measure the return and supply air temperatures.

   - Do not measure the temperature in an area where the thermometer can sense the radiant heat from the heat strips (see Figure 2-2). True air temperature cannot be measured if the thermometer senses radiant heat.

   - Take the temperature measurements within 6 ft of the air handler. Measurements taken at the return and supply grilles that are at too great a distance from the unit are not usually accurate enough.

   - Use the average temperature when more than one duct is connected to the supply air plenum.

   - Be sure the air temperature has stabilized before taking the temperature measurements.

   - Take the temperature measurements downstream from any source of mixed air. Use the following formula:

   $$\text{cfm} = \frac{\text{Btu}}{1.08 \times \Delta T}$$

   cfm = _____

7. Is this what the manufacturer rates the equipment? _____

# Electric Heating Unit Capacity Worksheet

Introduction: When a customer complains about insufficient heat from an electric heating unit, the service technician should, as a first step, determine if the heating elements are delivering the amount of heat they were designed to deliver. It could be that one or more of the heating elements is not operating properly. This is a fairly simple test and is easily performed.

Tools Needed: ammeter, voltmeter, accurate thermometer, and tool kit. An accurate wattmeter may be used in place of the ammeter and the voltmeter.

Procedures:

1. Turn off the electricity to any other equipment that is used in conjunction with the electric heating unit.

2. Set the thermostat to demand heat. Allow the heater to operate for approximately 10 minutes so the temperatures will stabilize.

3. Measure the voltage and amperage of each heating element and record:

|        | Volts | Amps |
|--------|-------|------|
| Motor: | 1._____ | _____ |
| Heater: | 1._____ | _____ |
|        | 2._____ | _____ |
|        | 3._____ | _____ |
|        | 4._____ | _____ |
|        | 5._____ | _____ |
| Sum:   | _____ | _____ |

4. Determine the total wattage of the heat strips and record. Use the following formula:

$$W = V \times I$$

W = _____

5. Determine the Btu output of the heat strips and record. Use the following formula:

$$Btu = W \times 3.413$$

Btu = _____

Copyright © Business News Publishing Company

6. Determine the cfm of the blower and record. Use the following instructions:

    - Use the same thermometer, or two that measure exactly the same, to measure the return and supply air temperatures.

    - Do not measure the temperature in an area where the thermometer can sense the radiant heat from the heat strips (see Figure 2-2). True air temperature cannot be measured if the thermometer senses radiant heat.

    - Take the temperature measurements within 6 ft of the air handler. Measurements taken at the return and supply grilles that are at too great a distance from the unit are not usually accurate enough.

    - Use the average temperature when more than one duct is connected to the supply air plenum.

    - Be sure the air temperature has stabilized before taking the temperature measurements.

    - Take the temperature measurements downstream from any source of mixed air. Use the following formula:

    $$\text{cfm} = \frac{\text{Btu}}{1.08 \times \Delta T}$$

    cfm = _____

7. Is this what the manufacturer rates the equipment? _____

# Electric Heating Unit Capacity Worksheet

Introduction: When a customer complains about insufficient heat from an electric heating unit, the service technician should, as a first step, determine if the heating elements are delivering the amount of heat they were designed to deliver. It could be that one or more of the heating elements is not operating properly. This is a fairly simple test and is easily performed.

Tools Needed: ammeter, voltmeter, accurate thermometer, and tool kit. An accurate wattmeter may be used in place of the ammeter and the voltmeter.

Procedures:

1. Turn off the electricity to any other equipment that is used in conjunction with the electric heating unit.

2. Set the thermostat to demand heat. Allow the heater to operate for approximately 10 minutes so the temperatures will stabilize.

3. Measure the voltage and amperage of each heating element and record:

|  | Volts | Amps |
|---|---|---|
| Motor: 1. | _____ | _____ |
| Heater: 1. | _____ | _____ |
| 2. | _____ | _____ |
| 3. | _____ | _____ |
| 4. | _____ | _____ |
| 5. | _____ | _____ |
| Sum: | _____ | _____ |

4. Determine the total wattage of the heat strips and record. Use the following formula:

$$W = V \times I$$

$W = $ _____

5. Determine the Btu output of the heat strips and record. Use the following formula:

$$Btu = W \times 3.413$$

$Btu = $ _____

Copyright © Business News Publishing Company

6. Determine the cfm of the blower and record. Use the following instructions:

    - Use the same thermometer, or two that measure exactly the same, to measure the return and supply air temperatures.

    - Do not measure the temperature in an area where the thermometer can sense the radiant heat from the heat strips (see Figure 2-2). True air temperature cannot be measured if the thermometer senses radiant heat.

    - Take the temperature measurements within 6 ft of the air handler. Measurements taken at the return and supply grilles that are at too great a distance from the unit are not usually accurate enough.

    - Use the average temperature when more than one duct is connected to the supply air plenum.

    - Be sure the air temperature has stabilized before taking the temperature measurements.

    - Take the temperature measurements downstream from any source of mixed air. Use the following formula:

    $$\text{cfm} = \frac{\text{Btu}}{1.08 \times \Delta T}$$

    cfm = _____

7. Is this what the manufacturer rates the equipment? _____

# Electric Heating Unit Capacity Worksheet

Introduction: When a customer complains about insufficient heat from an electric heating unit, the service technician should, as a first step, determine if the heating elements are delivering the amount of heat they were designed to deliver. It could be that one or more of the heating elements is not operating properly. This is a fairly simple test and is easily performed.

Tools Needed: ammeter, voltmeter, accurate thermometer, and tool kit. An accurate wattmeter may be used in place of the ammeter and the voltmeter.

Procedures:

1. Turn off the electricity to any other equipment that is used in conjunction with the electric heating unit.

2. Set the thermostat to demand heat. Allow the heater to operate for approximately 10 minutes so the temperatures will stabilize.

3. Measure the voltage and amperage of each heating element and record:

    Volts                    Amps

Motor:   1._____    _____

Heater:  1._____    _____

         2._____    _____

         3._____    _____

         4._____    _____

         5._____    _____

Sum: _____    _____


4. Determine the total wattage of the heat strips and record. Use the following formula:

    $W = V \times I$

    W = _____

5. Determine the Btu output of the heat strips and record. Use the following formula:

    $Btu = W \times 3.413$

    Btu = _____

Copyright © Business News Publishing Company

6. Determine the cfm of the blower and record. Use the following instructions:

   - Use the same thermometer, or two that measure exactly the same, to measure the return and supply air temperatures.

   - Do not measure the temperature in an area where the thermometer can sense the radiant heat from the heat strips (see Figure 2-2). True air temperature cannot be measured if the thermometer senses radiant heat.

   - Take the temperature measurements within 6 ft of the air handler. Measurements taken at the return and supply grilles that are at too great a distance from the unit are not usually accurate enough.

   - Use the average temperature when more than one duct is connected to the supply air plenum.

   - Be sure the air temperature has stabilized before taking the temperature measurements.

   - Take the temperature measurements downstream from any source of mixed air. Use the following formula:

   $$\text{cfm} = \frac{\text{Btu}}{1.08 \times \Delta T}$$

   cfm = _____

7. Is this what the manufacturer rates the equipment? _____

# Electric Heating Unit Capacity Worksheet

Introduction: When a customer complains about insufficient heat from an electric heating unit, the service technician should, as a first step, determine if the heating elements are delivering the amount of heat they were designed to deliver. It could be that one or more of the heating elements is not operating properly. This is a fairly simple test and is easily performed.

Tools Needed: ammeter, voltmeter, accurate thermometer, and tool kit. An accurate wattmeter may be used in place of the ammeter and the voltmeter.

Procedures:

1. Turn off the electricity to any other equipment that is used in conjunction with the electric heating unit.

2. Set the thermostat to demand heat. Allow the heater to operate for approximately 10 minutes so the temperatures will stabilize.

3. Measure the voltage and amperage of each heating element and record:

|  |  | Volts | Amps |
|---|---|---|---|
| Motor: | 1. | _____ | _____ |
| Heater: | 1. | _____ | _____ |
|  | 2. | _____ | _____ |
|  | 3. | _____ | _____ |
|  | 4. | _____ | _____ |
|  | 5. | _____ | _____ |
| Sum: |  | _____ | _____ |

4. Determine the total wattage of the heat strips and record. Use the following formula:

$$W = V \times I$$

W = _____

5. Determine the Btu output of the heat strips and record. Use the following formula:

$$Btu = W \times 3.413$$

Btu = _____

Copyright © Business News Publishing Company

6. Determine the cfm of the blower and record. Use the following instructions:

   - Use the same thermometer, or two that measure exactly the same, to measure the return and supply air temperatures.

   - Do not measure the temperature in an area where the thermometer can sense the radiant heat from the heat strips (see Figure 2-2). True air temperature cannot be measured if the thermometer senses radiant heat.

   - Take the temperature measurements within 6 ft of the air handler. Measurements taken at the return and supply grilles that are at too great a distance from the unit are not usually accurate enough.

   - Use the average temperature when more than one duct is connected to the supply air plenum.

   - Be sure the air temperature has stabilized before taking the temperature measurements.

   - Take the temperature measurements downstream from any source of mixed air. Use the following formula:

   $$cfm = \frac{Btu}{1.08 \times \Delta T}$$

   cfm = _____

7. Is this what the manufacturer rates the equipment? _____

# Electric Heating Unit Capacity Worksheet

Introduction: When a customer complains about insufficient heat from an electric heating unit, the service technician should, as a first step, determine if the heating elements are delivering the amount of heat they were designed to deliver. It could be that one or more of the heating elements is not operating properly. This is a fairly simple test and is easily performed.

Tools Needed: ammeter, voltmeter, accurate thermometer, and tool kit. An accurate wattmeter may be used in place of the ammeter and the voltmeter.

Procedures:

1. Turn off the electricity to any other equipment that is used in conjunction with the electric heating unit.

2. Set the thermostat to demand heat. Allow the heater to operate for approximately 10 minutes so the temperatures will stabilize.

3. Measure the voltage and amperage of each heating element and record:

    |        | Volts | Amps |
    |--------|-------|------|
    | Motor: 1. | _____ | _____ |
    | Heater: 1. | _____ | _____ |
    | 2. | _____ | _____ |
    | 3. | _____ | _____ |
    | 4. | _____ | _____ |
    | 5. | _____ | _____ |
    | Sum: | _____ | _____ |

4. Determine the total wattage of the heat strips and record. Use the following formula:

    $$W = V \times I$$

    W = _____

5. Determine the Btu output of the heat strips and record. Use the following formula:

    $$Btu = W \times 3.413$$

    Btu = _____

Copyright © Business News Publishing Company

6. Determine the cfm of the blower and record. Use the following instructions:

- Use the same thermometer, or two that measure exactly the same, to measure the return and supply air temperatures.

- Do not measure the temperature in an area where the thermometer can sense the radiant heat from the heat strips (see Figure 2-2). True air temperature cannot be measured if the thermometer senses radiant heat.

- Take the temperature measurements within 6 ft of the air handler. Measurements taken at the return and supply grilles that are at too great a distance from the unit are not usually accurate enough.

- Use the average temperature when more than one duct is connected to the supply air plenum.

- Be sure the air temperature has stabilized before taking the temperature measurements.

- Take the temperature measurements downstream from any source of mixed air. Use the following formula:

$$\text{cfm} = \frac{\text{Btu}}{1.08 \times \Delta T}$$

cfm = _____

7. Is this what the manufacturer rates the equipment? _____

# Electric Heating Unit Capacity Worksheet

Introduction: When a customer complains about insufficient heat from an electric heating unit, the service technician should, as a first step, determine if the heating elements are delivering the amount of heat they were designed to deliver. It could be that one or more of the heating elements is not operating properly. This is a fairly simple test and is easily performed.

Tools Needed: ammeter, voltmeter, accurate thermometer, and tool kit. An accurate wattmeter may be used in place of the ammeter and the voltmeter.

Procedures:

1. Turn off the electricity to any other equipment that is used in conjunction with the electric heating unit.

2. Set the thermostat to demand heat. Allow the heater to operate for approximately 10 minutes so the temperatures will stabilize.

3. Measure the voltage and amperage of each heating element and record:

|  | Volts | Amps |
|---|---|---|
| Motor: | 1._____ | _____ |
| Heater: | 1._____ | _____ |
|  | 2._____ | _____ |
|  | 3._____ | _____ |
|  | 4._____ | _____ |
|  | 5._____ | _____ |
| Sum: | _____ | _____ |

4. Determine the total wattage of the heat strips and record. Use the following formula:

$$W = V \times I$$

W = _____

5. Determine the Btu output of the heat strips and record. Use the following formula:

$$Btu = W \times 3.413$$

Btu = _____

6. Determine the cfm of the blower and record. Use the following instructions:

    - Use the same thermometer, or two that measure exactly the same, to measure the return and supply air temperatures.

    - Do not measure the temperature in an area where the thermometer can sense the radiant heat from the heat strips (see Figure 2-2). True air temperature cannot be measured if the thermometer senses radiant heat.

    - Take the temperature measurements within 6 ft of the air handler. Measurements taken at the return and supply grilles that are at too great a distance from the unit are not usually accurate enough.

    - Use the average temperature when more than one duct is connected to the supply air plenum.

    - Be sure the air temperature has stabilized before taking the temperature measurements.

    - Take the temperature measurements downstream from any source of mixed air. Use the following formula:

    $$\text{cfm} = \frac{\text{Btu}}{1.08 \times \Delta T}$$

    cfm = _____

7. Is this what the manufacturer rates the equipment?_____

# Electric Heating Unit Capacity Worksheet

Introduction: When a customer complains about insufficient heat from an electric heating unit, the service technician should, as a first step, determine if the heating elements are delivering the amount of heat they were designed to deliver. It could be that one or more of the heating elements is not operating properly. This is a fairly simple test and is easily performed.

Tools Needed: ammeter, voltmeter, accurate thermometer, and tool kit. An accurate wattmeter may be used in place of the ammeter and the voltmeter.

Procedures:

1. Turn off the electricity to any other equipment that is used in conjunction with the electric heating unit.

2. Set the thermostat to demand heat. Allow the heater to operate for approximately 10 minutes so the temperatures will stabilize.

3. Measure the voltage and amperage of each heating element and record:

|  | Volts | Amps |
|---|---|---|
| Motor: 1. | _____ | _____ |
| Heater: 1. | _____ | _____ |
| 2. | _____ | _____ |
| 3. | _____ | _____ |
| 4. | _____ | _____ |
| 5. | _____ | _____ |
| Sum: | _____ | _____ |

4. Determine the total wattage of the heat strips and record. Use the following formula:

$$W = V \times I$$

$W = $ _____

5. Determine the Btu output of the heat strips and record. Use the following formula:

$$Btu = W \times 3.413$$

$Btu = $ _____

Copyright © Business News Publishing Company

6. Determine the cfm of the blower and record. Use the following instructions:

    - Use the same thermometer, or two that measure exactly the same, to measure the return and supply air temperatures.

    - Do not measure the temperature in an area where the thermometer can sense the radiant heat from the heat strips (see Figure 2-2). True air temperature cannot be measured if the thermometer senses radiant heat.

    - Take the temperature measurements within 6 ft of the air handler. Measurements taken at the return and supply grilles that are at too great a distance from the unit are not usually accurate enough.

    - Use the average temperature when more than one duct is connected to the supply air plenum.

    - Be sure the air temperature has stabilized before taking the temperature measurements.

    - Take the temperature measurements downstream from any source of mixed air. Use the following formula:

    $$cfm = \frac{Btu}{1.08 \times \Delta T}$$

    cfm = _____

7. Is this what the manufacturer rates the equipment? _____

# Gas Heating Worksheet

Introduction: Use the following procedures and the test instrument manufacturers' instructions to determine the combustion efficiency of gas burning equipment.

Tools Needed: tool kit, gas manifold pressure gauge, dry bulb thermometer, flue gas temperature thermometer, draft gauge, carbon monoxide analyzer, carbon dioxide analyzer, and velometer.

Procedures:

1. Visually check the entire system for cleanliness, and ensure all components are in proper working condition. Be sure the heat exchanger passages and the venting system are clear of all obstructions.

2. Determine the type of gas (natural, LP) and record. _____

3. Determine the Btu content of the gas and record. _____ per cubic foot

4. Start the heating unit, and allow it to operate for about ten minutes.

5. Measure the manifold gas pressure and record. _____ inches w.c.

6. Determine the type of flame and record. _____

7. Adjust the burner if needed, and record the type of flame. _____

8. Determine the temperature rise of the circulating air through the unit and record. Use the following formula:

    $\Delta T$ = Discharge air temperature - Entering air temperature

    $\Delta T$ = _____ °F

9. Determine the blower cfm and record. Use the following formula:

    $$cfm = \frac{Btu}{1.08 \times \Delta T}$$

    cfm = _____

10. Check the flue gas temperature and record. _____ °F

11. Check the $CO_2$ content of the flue gases and record. _____ %

12. Determine the operating combustion efficiency and record. _____ %

13. Measure the unit vent draft and record. _____ inches w.c.

14. Measure the CO content of the flue gases and record. _____ %

15. Make any adjustments or repairs required to increase the combustion efficiency of the unit. Repeat Steps 3 through 13.

16. Is this what the manufacturer rates the unit? _____

Copyright © Business News Publishing Company

# Gas Heating Worksheet

Introduction: Use the following procedures and the test instrument manufacturers' instructions to determine the combustion efficiency of gas burning equipment.

Tools Needed: tool kit, gas manifold pressure gauge, dry bulb thermometer, flue gas temperature thermometer, draft gauge, carbon monoxide analyzer, carbon dioxide analyzer, and velometer.

Procedures:

1. Visually check the entire system for cleanliness, and ensure all components are in proper working condition. Be sure the heat exchanger passages and the venting system are clear of all obstructions.

2. Determine the type of gas (natural, LP) and record. _____

3. Determine the Btu content of the gas and record. _____ per cubic foot

4. Start the heating unit, and allow it to operate for about ten minutes.

5. Measure the manifold gas pressure and record. _____ inches w.c.

6. Determine the type of flame and record. _____

7. Adjust the burner if needed, and record the type of flame. _____

8. Determine the temperature rise of the circulating air through the unit and record. Use the following formula:

    $\Delta T$ = Discharge air temperature - Entering air temperature

    $\Delta T$ = _____ °F

9. Determine the blower cfm and record. Use the following formula:

    $$cfm = \frac{Btu}{1.08 \times \Delta T}$$

    cfm = _____

10. Check the flue gas temperature and record. _____ °F

11. Check the $CO_2$ content of the flue gases and record. _____ %

12. Determine the operating combustion efficiency and record. _____ %

13. Measure the unit vent draft and record. _____ inches w.c.

14. Measure the CO content of the flue gases and record. _____ %

15. Make any adjustments or repairs required to increase the combustion efficiency of the unit. Repeat Steps 3 through 13.

16. Is this what the manufacturer rates the unit? _____

Copyright © Business News Publishing Company

# Gas Heating Worksheet

Introduction: Use the following procedures and the test instrument manufacturers' instructions to determine the combustion efficiency of gas burning equipment.

Tools Needed: tool kit, gas manifold pressure gauge, dry bulb thermometer, flue gas temperature thermometer, draft gauge, carbon monoxide analyzer, carbon dioxide analyzer, and velometer.

Procedures:

1. Visually check the entire system for cleanliness, and ensure all components are in proper working condition. Be sure the heat exchanger passages and the venting system are clear of all obstructions.

2. Determine the type of gas (natural, LP) and record. _____

3. Determine the Btu content of the gas and record. _____ per cubic foot

4. Start the heating unit, and allow it to operate for about ten minutes.

5. Measure the manifold gas pressure and record. _____ inches w.c.

6. Determine the type of flame and record. _____

7. Adjust the burner if needed, and record the type of flame. _____

8. Determine the temperature rise of the circulating air through the unit and record. Use the following formula:

$$\Delta T = \text{Discharge air temperature} - \text{Entering air temperature}$$

$\Delta T =$ _____ °F

9. Determine the blower cfm and record. Use the following formula:

$$\text{cfm} = \frac{\text{Btu}}{1.08 \times \Delta T}$$

cfm = _____

10. Check the flue gas temperature and record. _____ °F

11. Check the $CO_2$ content of the flue gases and record. _____ %

12. Determine the operating combustion efficiency and record. _____ %

13. Measure the unit vent draft and record. _____ inches w.c.

14. Measure the CO content of the flue gases and record. _____ %

15. Make any adjustments or repairs required to increase the combustion efficiency of the unit. Repeat Steps 3 through 13.

16. Is this what the manufacturer rates the unit? _____

Copyright © Business News Publishing Company

# Gas Heating Worksheet

Introduction: Use the following procedures and the test instrument manufacturers' instructions to determine the combustion efficiency of gas burning equipment.

Tools Needed: tool kit, gas manifold pressure gauge, dry bulb thermometer, flue gas temperature thermometer, draft gauge, carbon monoxide analyzer, carbon dioxide analyzer, and velometer.

Procedures:

1. Visually check the entire system for cleanliness, and ensure all components are in proper working condition. Be sure the heat exchanger passages and the venting system are clear of all obstructions.

2. Determine the type of gas (natural, LP) and record. _____

3. Determine the Btu content of the gas and record. _____ per cubic foot

4. Start the heating unit, and allow it to operate for about ten minutes.

5. Measure the manifold gas pressure and record. _____ inches w.c.

6. Determine the type of flame and record. _____

7. Adjust the burner if needed, and record the type of flame. _____

8. Determine the temperature rise of the circulating air through the unit and record. Use the following formula:

    $\Delta T$ = Discharge air temperature - Entering air temperature

    $\Delta T$ = _____ °F

9. Determine the blower cfm and record. Use the following formula:

    $$cfm = \frac{Btu}{1.08 \times \Delta T}$$

    cfm = _____

10. Check the flue gas temperature and record. _____ °F

11. Check the $CO_2$ content of the flue gases and record. _____ %

12. Determine the operating combustion efficiency and record. _____ %

13. Measure the unit vent draft and record. _____ inches w.c.

14. Measure the CO content of the flue gases and record. _____ %

15. Make any adjustments or repairs required to increase the combustion efficiency of the unit. Repeat Steps 3 through 13.

16. Is this what the manufacturer rates the unit? _____

Copyright © Business News Publishing Company

# Gas Heating Worksheet

Introduction: Use the following procedures and the test instrument manufacturers' instructions to determine the combustion efficiency of gas burning equipment.

Tools Needed: tool kit, gas manifold pressure gauge, dry bulb thermometer, flue gas temperature thermometer, draft gauge, carbon monoxide analyzer, carbon dioxide analyzer, and velometer.

Procedures:

1. Visually check the entire system for cleanliness, and ensure all components are in proper working condition. Be sure the heat exchanger passages and the venting system are clear of all obstructions.

2. Determine the type of gas (natural, LP) and record. _____

3. Determine the Btu content of the gas and record. _____ per cubic foot

4. Start the heating unit, and allow it to operate for about ten minutes.

5. Measure the manifold gas pressure and record. _____ inches w.c.

6. Determine the type of flame and record. _____

7. Adjust the burner if needed, and record the type of flame. _____

8. Determine the temperature rise of the circulating air through the unit and record. Use the following formula:

    $\Delta T$ = Discharge air temperature - Entering air temperature

    $\Delta T$ = _____ °F

9. Determine the blower cfm and record. Use the following formula:

    $$cfm = \frac{Btu}{1.08 \times \Delta T}$$

    cfm = _____

10. Check the flue gas temperature and record. _____ °F

11. Check the $CO_2$ content of the flue gases and record. _____ %

12. Determine the operating combustion efficiency and record. _____ %

13. Measure the unit vent draft and record. _____ inches w.c.

14. Measure the CO content of the flue gases and record. _____ %

15. Make any adjustments or repairs required to increase the combustion efficiency of the unit. Repeat Steps 3 through 13.

16. Is this what the manufacturer rates the unit? _____

Copyright © Business News Publishing Company

# Gas Heating Worksheet

Introduction: Use the following procedures and the test instrument manufacturers' instructions to determine the combustion efficiency of gas burning equipment.

Tools Needed: tool kit, gas manifold pressure gauge, dry bulb thermometer, flue gas temperature thermometer, draft gauge, carbon monoxide analyzer, carbon dioxide analyzer, and velometer.

Procedures:

1. Visually check the entire system for cleanliness, and ensure all components are in proper working condition. Be sure the heat exchanger passages and the venting system are clear of all obstructions.

2. Determine the type of gas (natural, LP) and record. _____

3. Determine the Btu content of the gas and record. _____ per cubic foot

4. Start the heating unit, and allow it to operate for about ten minutes.

5. Measure the manifold gas pressure and record. _____ inches w.c.

6. Determine the type of flame and record. _____

7. Adjust the burner if needed, and record the type of flame. _____

8. Determine the temperature rise of the circulating air through the unit and record. Use the following formula:

    $\Delta T$ = Discharge air temperature - Entering air temperature

    $\Delta T$ = _____ °F

9. Determine the blower cfm and record. Use the following formula:

    $$\text{cfm} = \frac{\text{Btu}}{1.08 \times \Delta T}$$

    cfm = _____

10. Check the flue gas temperature and record. _____ °F

11. Check the $CO_2$ content of the flue gases and record. _____ %

12. Determine the operating combustion efficiency and record. _____ %

13. Measure the unit vent draft and record. _____ inches w.c.

14. Measure the CO content of the flue gases and record. _____ %

15. Make any adjustments or repairs required to increase the combustion efficiency of the unit. Repeat Steps 3 through 13.

16. Is this what the manufacturer rates the unit? _____

Copyright © Business News Publishing Company

# Gas Heating Worksheet

Introduction: Use the following procedures and the test instrument manufacturers' instructions to determine the combustion efficiency of gas burning equipment.

Tools Needed: tool kit, gas manifold pressure gauge, dry bulb thermometer, flue gas temperature thermometer, draft gauge, carbon monoxide analyzer, carbon dioxide analyzer, and velometer.

Procedures:

1. Visually check the entire system for cleanliness, and ensure all components are in proper working condition. Be sure the heat exchanger passages and the venting system are clear of all obstructions.

2. Determine the type of gas (natural, LP) and record. _____

3. Determine the Btu content of the gas and record. _____ per cubic foot

4. Start the heating unit, and allow it to operate for about ten minutes.

5. Measure the manifold gas pressure and record. _____ inches w.c.

6. Determine the type of flame and record. _____

7. Adjust the burner if needed, and record the type of flame. _____

8. Determine the temperature rise of the circulating air through the unit and record. Use the following formula:

    $\Delta T$ = Discharge air temperature - Entering air temperature

    $\Delta T$ = _____ °F

9. Determine the blower cfm and record. Use the following formula:

    $$cfm = \frac{Btu}{1.08 \times \Delta T}$$

    cfm = _____

10. Check the flue gas temperature and record. _____ °F

11. Check the $CO_2$ content of the flue gases and record. _____ %

12. Determine the operating combustion efficiency and record. _____ %

13. Measure the unit vent draft and record. _____ inches w.c.

14. Measure the CO content of the flue gases and record. _____ %

15. Make any adjustments or repairs required to increase the combustion efficiency of the unit. Repeat Steps 3 through 13.

16. Is this what the manufacturer rates the unit? _____

Copyright © Business News Publishing Company

# Gas Heating Worksheet

Introduction: Use the following procedures and the test instrument manufacturers' instructions to determine the combustion efficiency of gas burning equipment.

Tools Needed: tool kit, gas manifold pressure gauge, dry bulb thermometer, flue gas temperature thermometer, draft gauge, carbon monoxide analyzer, carbon dioxide analyzer, and velometer.

Procedures:

1. Visually check the entire system for cleanliness, and ensure all components are in proper working condition. Be sure the heat exchanger passages and the venting system are clear of all obstructions.

2. Determine the type of gas (natural, LP) and record. _____

3. Determine the Btu content of the gas and record. _____ per cubic foot

4. Start the heating unit, and allow it to operate for about ten minutes.

5. Measure the manifold gas pressure and record. _____ inches w.c.

6. Determine the type of flame and record. _____

7. Adjust the burner if needed, and record the type of flame. _____

8. Determine the temperature rise of the circulating air through the unit and record. Use the following formula:

    $\Delta T$ = Discharge air temperature - Entering air temperature

    $\Delta T$ = _____ °F

9. Determine the blower cfm and record. Use the following formula:

    $$cfm = \frac{Btu}{1.08 \times \Delta T}$$

    cfm = _____

10. Check the flue gas temperature and record. _____ °F

11. Check the $CO_2$ content of the flue gases and record. _____ %

12. Determine the operating combustion efficiency and record. _____ %

13. Measure the unit vent draft and record. _____ inches w.c.

14. Measure the CO content of the flue gases and record. _____ %

15. Make any adjustments or repairs required to increase the combustion efficiency of the unit. Repeat Steps 3 through 13.

16. Is this what the manufacturer rates the unit? _____

Copyright © Business News Publishing Company

# Gas Heating Worksheet

Introduction: Use the following procedures and the test instrument manufacturers' instructions to determine the combustion efficiency of gas burning equipment.

Tools Needed: tool kit, gas manifold pressure gauge, dry bulb thermometer, flue gas temperature thermometer, draft gauge, carbon monoxide analyzer, carbon dioxide analyzer, and velometer.

Procedures:

1. Visually check the entire system for cleanliness, and ensure all components are in proper working condition. Be sure the heat exchanger passages and the venting system are clear of all obstructions.

2. Determine the type of gas (natural, LP) and record. _____

3. Determine the Btu content of the gas and record. _____ per cubic foot

4. Start the heating unit, and allow it to operate for about ten minutes.

5. Measure the manifold gas pressure and record. _____ inches w.c.

6. Determine the type of flame and record. _____

7. Adjust the burner if needed, and record the type of flame. _____

8. Determine the temperature rise of the circulating air through the unit and record. Use the following formula:

    $\Delta T$ = Discharge air temperature - Entering air temperature

    $\Delta T$ = _____ °F

9. Determine the blower cfm and record. Use the following formula:

    $$cfm = \frac{Btu}{1.08 \times \Delta T}$$

    cfm = _____

10. Check the flue gas temperature and record. _____ °F

11. Check the $CO_2$ content of the flue gases and record. _____ %

12. Determine the operating combustion efficiency and record. _____ %

13. Measure the unit vent draft and record. _____ inches w.c.

14. Measure the CO content of the flue gases and record. _____ %

15. Make any adjustments or repairs required to increase the combustion efficiency of the unit. Repeat Steps 3 through 13.

16. Is this what the manufacturer rates the unit? _____

Copyright © Business News Publishing Company

# Gas Heating Worksheet

Introduction: Use the following procedures and the test instrument manufacturers' instructions to determine the combustion efficiency of gas burning equipment.

Tools Needed: tool kit, gas manifold pressure gauge, dry bulb thermometer, flue gas temperature thermometer, draft gauge, carbon monoxide analyzer, carbon dioxide analyzer, and velometer.

Procedures:

1. Visually check the entire system for cleanliness, and ensure all components are in proper working condition. Be sure the heat exchanger passages and the venting system are clear of all obstructions.

2. Determine the type of gas (natural, LP) and record. _____

3. Determine the Btu content of the gas and record. _____ per cubic foot

4. Start the heating unit, and allow it to operate for about ten minutes.

5. Measure the manifold gas pressure and record. _____ inches w.c.

6. Determine the type of flame and record. _____

7. Adjust the burner if needed, and record the type of flame. _____

8. Determine the temperature rise of the circulating air through the unit and record. Use the following formula:

    $\Delta T$ = Discharge air temperature - Entering air temperature

    $\Delta T$ = _____ °F

9. Determine the blower cfm and record. Use the following formula:

    $$\text{cfm} = \frac{\text{Btu}}{1.08 \times \Delta T}$$

    cfm = _____

10. Check the flue gas temperature and record. _____ °F

11. Check the $CO_2$ content of the flue gases and record. _____ %

12. Determine the operating combustion efficiency and record. _____ %

13. Measure the unit vent draft and record. _____ inches w.c.

14. Measure the CO content of the flue gases and record. _____ %

15. Make any adjustments or repairs required to increase the combustion efficiency of the unit. Repeat Steps 3 through 13.

16. Is this what the manufacturer rates the unit? _____

Copyright © Business News Publishing Company

# Gas Heating Worksheet

Introduction: Use the following procedures and the test instrument manufacturers' instructions to determine the combustion efficiency of gas burning equipment.

Tools Needed: tool kit, gas manifold pressure gauge, dry bulb thermometer, flue gas temperature thermometer, draft gauge, carbon monoxide analyzer, carbon dioxide analyzer, and velometer.

Procedures:

1. Visually check the entire system for cleanliness, and ensure all components are in proper working condition. Be sure the heat exchanger passages and the venting system are clear of all obstructions.

2. Determine the type of gas (natural, LP) and record. _____

3. Determine the Btu content of the gas and record. _____ per cubic foot

4. Start the heating unit, and allow it to operate for about ten minutes.

5. Measure the manifold gas pressure and record. _____ inches w.c.

6. Determine the type of flame and record. _____

7. Adjust the burner if needed, and record the type of flame. _____

8. Determine the temperature rise of the circulating air through the unit and record. Use the following formula:

$$\Delta T = \text{Discharge air temperature - Entering air temperature}$$

$\Delta T = $ _____ °F

9. Determine the blower cfm and record. Use the following formula:

$$\text{cfm} = \frac{\text{Btu}}{1.08 \times \Delta T}$$

cfm = _____

10. Check the flue gas temperature and record. _____ °F

11. Check the $CO_2$ content of the flue gases and record. _____ %

12. Determine the operating combustion efficiency and record. _____ %

13. Measure the unit vent draft and record. _____ inches w.c.

14. Measure the CO content of the flue gases and record. _____ %

15. Make any adjustments or repairs required to increase the combustion efficiency of the unit. Repeat Steps 3 through 13.

16. Is this what the manufacturer rates the unit? _____

Copyright © Business News Publishing Company

# Gas Heating Worksheet

Introduction: Use the following procedures and the test instrument manufacturers' instructions to determine the combustion efficiency of gas burning equipment.

Tools Needed: tool kit, gas manifold pressure gauge, dry bulb thermometer, flue gas temperature thermometer, draft gauge, carbon monoxide analyzer, carbon dioxide analyzer, and velometer.

Procedures:

1. Visually check the entire system for cleanliness, and ensure all components are in proper working condition. Be sure the heat exchanger passages and the venting system are clear of all obstructions.

2. Determine the type of gas (natural, LP) and record. _____

3. Determine the Btu content of the gas and record. _____ per cubic foot

4. Start the heating unit, and allow it to operate for about ten minutes.

5. Measure the manifold gas pressure and record. _____ inches w.c.

6. Determine the type of flame and record. _____

7. Adjust the burner if needed, and record the type of flame. _____

8. Determine the temperature rise of the circulating air through the unit and record. Use the following formula:

    $\Delta T$ = Discharge air temperature - Entering air temperature

    $\Delta T$ = _____ °F

9. Determine the blower cfm and record. Use the following formula:

    $$cfm = \frac{Btu}{1.08 \times \Delta T}$$

    cfm = _____

10. Check the flue gas temperature and record. _____ °F

11. Check the $CO_2$ content of the flue gases and record. _____ %

12. Determine the operating combustion efficiency and record. _____ %

13. Measure the unit vent draft and record. _____ inches w.c.

14. Measure the CO content of the flue gases and record. _____ %

15. Make any adjustments or repairs required to increase the combustion efficiency of the unit. Repeat Steps 3 through 13.

16. Is this what the manufacturer rates the unit? _____

Copyright © Business News Publishing Company

# Gas Heating Worksheet

Introduction: Use the following procedures and the test instrument manufacturers' instructions to determine the combustion efficiency of gas burning equipment.

Tools Needed: tool kit, gas manifold pressure gauge, dry bulb thermometer, flue gas temperature thermometer, draft gauge, carbon monoxide analyzer, carbon dioxide analyzer, and velometer.

Procedures:

1. Visually check the entire system for cleanliness, and ensure all components are in proper working condition. Be sure the heat exchanger passages and the venting system are clear of all obstructions.

2. Determine the type of gas (natural, LP) and record. _____

3. Determine the Btu content of the gas and record. _____ per cubic foot

4. Start the heating unit, and allow it to operate for about ten minutes.

5. Measure the manifold gas pressure and record. _____ inches w.c.

6. Determine the type of flame and record. _____

7. Adjust the burner if needed, and record the type of flame. _____

8. Determine the temperature rise of the circulating air through the unit and record. Use the following formula:

    $\Delta T$ = Discharge air temperature - Entering air temperature

    $\Delta T$ = _____ °F

9. Determine the blower cfm and record. Use the following formula:

    $$cfm = \frac{Btu}{1.08 \times \Delta T}$$

    cfm = _____

10. Check the flue gas temperature and record. _____ °F

11. Check the $CO_2$ content of the flue gases and record. _____ %

12. Determine the operating combustion efficiency and record. _____ %

13. Measure the unit vent draft and record. _____ inches w.c.

14. Measure the CO content of the flue gases and record. _____ %

15. Make any adjustments or repairs required to increase the combustion efficiency of the unit. Repeat Steps 3 through 13.

16. Is this what the manufacturer rates the unit? _____

Copyright © Business News Publishing Company

# Gas Heating Worksheet

Introduction: Use the following procedures and the test instrument manufacturers' instructions to determine the combustion efficiency of gas burning equipment.

Tools Needed: tool kit, gas manifold pressure gauge, dry bulb thermometer, flue gas temperature thermometer, draft gauge, carbon monoxide analyzer, carbon dioxide analyzer, and velometer.

Procedures:

1. Visually check the entire system for cleanliness, and ensure all components are in proper working condition. Be sure the heat exchanger passages and the venting system are clear of all obstructions.

2. Determine the type of gas (natural, LP) and record. _____

3. Determine the Btu content of the gas and record. _____ per cubic foot

4. Start the heating unit, and allow it to operate for about ten minutes.

5. Measure the manifold gas pressure and record. _____ inches w.c.

6. Determine the type of flame and record. _____

7. Adjust the burner if needed, and record the type of flame. _____

8. Determine the temperature rise of the circulating air through the unit and record. Use the following formula:

    $\Delta T$ = Discharge air temperature - Entering air temperature

    $\Delta T$ = _____ °F

9. Determine the blower cfm and record. Use the following formula:

    $$cfm = \frac{Btu}{1.08 \times \Delta T}$$

    cfm = _____

10. Check the flue gas temperature and record. _____ °F

11. Check the $CO_2$ content of the flue gases and record. _____ %

12. Determine the operating combustion efficiency and record. _____ %

13. Measure the unit vent draft and record. _____ inches w.c.

14. Measure the CO content of the flue gases and record. _____ %

15. Make any adjustments or repairs required to increase the combustion efficiency of the unit. Repeat Steps 3 through 13.

16. Is this what the manufacturer rates the unit? _____

Copyright © Business News Publishing Company

# Gas Heating Worksheet

Introduction: Use the following procedures and the test instrument manufacturers' instructions to determine the combustion efficiency of gas burning equipment.

Tools Needed: tool kit, gas manifold pressure gauge, dry bulb thermometer, flue gas temperature thermometer, draft gauge, carbon monoxide analyzer, carbon dioxide analyzer, and velometer.

Procedures:

1. Visually check the entire system for cleanliness, and ensure all components are in proper working condition. Be sure the heat exchanger passages and the venting system are clear of all obstructions.

2. Determine the type of gas (natural, LP) and record. _____

3. Determine the Btu content of the gas and record. _____ per cubic foot

4. Start the heating unit, and allow it to operate for about ten minutes.

5. Measure the manifold gas pressure and record. _____ inches w.c.

6. Determine the type of flame and record. _____

7. Adjust the burner if needed, and record the type of flame. _____

8. Determine the temperature rise of the circulating air through the unit and record. Use the following formula:

    $\Delta T$ = Discharge air temperature - Entering air temperature

    $\Delta T$ = _____ °F

9. Determine the blower cfm and record. Use the following formula:

    $$\text{cfm} = \frac{\text{Btu}}{1.08 \times \Delta T}$$

    cfm = _____

10. Check the flue gas temperature and record. _____ °F

11. Check the $CO_2$ content of the flue gases and record. _____ %

12. Determine the operating combustion efficiency and record. _____ %

13. Measure the unit vent draft and record. _____ inches w.c.

14. Measure the CO content of the flue gases and record. _____ %

15. Make any adjustments or repairs required to increase the combustion efficiency of the unit. Repeat Steps 3 through 13.

16. Is this what the manufacturer rates the unit? _____

Copyright © Business News Publishing Company

# Oil Heating Worksheet

Introduction: When a customer complains about not enough heat from an oil-fired unit, the service technician should, as a first step, determine if the oil burner is delivering the amount of oil it was designed to deliver. Use the following procedures and the test instrument manufacturers' instructions to determine the combustion efficiency of oil burning equipment.

Tools Needed: tool kit, flue gas temperature thermometer, draft gauge, smoke tester, and carbon dioxide analyzer with the appropriate combustion efficiency chart or slide rule to use in combination with the various test results to determine the combustion efficiency.

Procedures:

1. Visually check the entire unit, and ensure all components are in proper working condition. Be sure the flue gas passages and the venting system are clear of all obstructions.

2. Is this a conversion burner? _____

3. Start the oil burner, and allow it to operate for about 15 minutes.

4. Check the flue gas temperature and record. _____°F

5. Check the $CO_2$ content of the flue gases and record. _____%

6. The operating combustion efficiency is _____%

7. Conduct the smoke test and record. _____spot

8. Measure the over-fire draft. _____inches w.c.

9. Make any adjustments or repairs required to increase the combustion efficiency of the unit. Repeat Steps 3 through 8.

10. Measure the unit cfm and record. _____cfm

11. Is this what the manufacturer rates the equipment? _____

# Oil Heating Worksheet

Introduction: When a customer complains about not enough heat from an oil-fired unit, the service technician should, as a first step, determine if the oil burner is delivering the amount of oil it was designed to deliver. Use the following procedures and the test instrument manufacturers' instructions to determine the combustion efficiency of oil burning equipment.

Tools Needed: tool kit, flue gas temperature thermometer, draft gauge, smoke tester, and carbon dioxide analyzer with the appropriate combustion efficiency chart or slide rule to use in combination with the various test results to determine the combustion efficiency.

Procedures:

1. Visually check the entire unit, and ensure all components are in proper working condition. Be sure the flue gas passages and the venting system are clear of all obstructions.

2. Is this a conversion burner? _____

3. Start the oil burner, and allow it to operate for about 15 minutes.

4. Check the flue gas temperature and record. _____ °F

5. Check the $CO_2$ content of the flue gases and record. _____ %

6. The operating combustion efficiency is _____ %

7. Conduct the smoke test and record. _____ spot

8. Measure the over-fire draft. _____ inches w.c.

9. Make any adjustments or repairs required to increase the combustion efficiency of the unit. Repeat Steps 3 through 8.

10. Measure the unit cfm and record. _____ cfm

11. Is this what the manufacturer rates the equipment? _____

Copyright © Business News Publishing Company

# Oil Heating Worksheet

Introduction: When a customer complains about not enough heat from an oil-fired unit, the service technician should, as a first step, determine if the oil burner is delivering the amount of oil it was designed to deliver. Use the following procedures and the test instrument manufacturers' instructions to determine the combustion efficiency of oil burning equipment.

Tools Needed: tool kit, flue gas temperature thermometer, draft gauge, smoke tester, and carbon dioxide analyzer with the appropriate combustion efficiency chart or slide rule to use in combination with the various test results to determine the combustion efficiency.

Procedures:

1. Visually check the entire unit, and ensure all components are in proper working condition. Be sure the flue gas passages and the venting system are clear of all obstructions.

2. Is this a conversion burner? _____

3. Start the oil burner, and allow it to operate for about 15 minutes.

4. Check the flue gas temperature and record. _____°F

5. Check the $CO_2$ content of the flue gases and record. _____%

6. The operating combustion efficiency is _____%

7. Conduct the smoke test and record. _____spot

8. Measure the over-fire draft. _____inches w.c.

9. Make any adjustments or repairs required to increase the combustion efficiency of the unit. Repeat Steps 3 through 8.

10. Measure the unit cfm and record. _____cfm

11. Is this what the manufacturer rates the equipment? _____

# Oil Heating Worksheet

Introduction: When a customer complains about not enough heat from an oil-fired unit, the service technician should, as a first step, determine if the oil burner is delivering the amount of oil it was designed to deliver. Use the following procedures and the test instrument manufacturers' instructions to determine the combustion efficiency of oil burning equipment.

Tools Needed: tool kit, flue gas temperature thermometer, draft gauge, smoke tester, and carbon dioxide analyzer with the appropriate combustion efficiency chart or slide rule to use in combination with the various test results to determine the combustion efficiency.

Procedures:

1. Visually check the entire unit, and ensure all components are in proper working condition. Be sure the flue gas passages and the venting system are clear of all obstructions.

2. Is this a conversion burner? _____

3. Start the oil burner, and allow it to operate for about 15 minutes.

4. Check the flue gas temperature and record. _____°F

5. Check the $CO_2$ content of the flue gases and record. _____%

6. The operating combustion efficiency is _____%

7. Conduct the smoke test and record. _____spot

8. Measure the over-fire draft. _____inches w.c.

9. Make any adjustments or repairs required to increase the combustion efficiency of the unit. Repeat Steps 3 through 8.

10. Measure the unit cfm and record. _____cfm

11. Is this what the manufacturer rates the equipment? _____

# Oil Heating Worksheet

Introduction: When a customer complains about not enough heat from an oil-fired unit, the service technician should, as a first step, determine if the oil burner is delivering the amount of oil it was designed to deliver. Use the following procedures and the test instrument manufacturers' instructions to determine the combustion efficiency of oil burning equipment.

Tools Needed: tool kit, flue gas temperature thermometer, draft gauge, smoke tester, and carbon dioxide analyzer with the appropriate combustion efficiency chart or slide rule to use in combination with the various test results to determine the combustion efficiency.

Procedures:

1. Visually check the entire unit, and ensure all components are in proper working condition. Be sure the flue gas passages and the venting system are clear of all obstructions.

2. Is this a conversion burner? _____

3. Start the oil burner, and allow it to operate for about 15 minutes.

4. Check the flue gas temperature and record. _____°F

5. Check the $CO_2$ content of the flue gases and record. _____%

6. The operating combustion efficiency is _____%

7. Conduct the smoke test and record. _____spot

8. Measure the over-fire draft. _____inches w.c.

9. Make any adjustments or repairs required to increase the combustion efficiency of the unit. Repeat Steps 3 through 8.

10. Measure the unit cfm and record. _____cfm

11. Is this what the manufacturer rates the equipment? _____

Copyright © Business News Publishing Company

# Oil Heating Worksheet

Introduction: When a customer complains about not enough heat from an oil-fired unit, the service technician should, as a first step, determine if the oil burner is delivering the amount of oil it was designed to deliver. Use the following procedures and the test instrument manufacturers' instructions to determine the combustion efficiency of oil burning equipment.

Tools Needed: tool kit, flue gas temperature thermometer, draft gauge, smoke tester, and carbon dioxide analyzer with the appropriate combustion efficiency chart or slide rule to use in combination with the various test results to determine the combustion efficiency.

Procedures:

1. Visually check the entire unit, and ensure all components are in proper working condition. Be sure the flue gas passages and the venting system are clear of all obstructions.

2. Is this a conversion burner? _____

3. Start the oil burner, and allow it to operate for about 15 minutes.

4. Check the flue gas temperature and record. _____ °F

5. Check the $CO_2$ content of the flue gases and record. _____ %

6. The operating combustion efficiency is _____ %

7. Conduct the smoke test and record. _____ spot

8. Measure the over-fire draft. _____ inches w.c.

9. Make any adjustments or repairs required to increase the combustion efficiency of the unit. Repeat Steps 3 through 8.

10. Measure the unit cfm and record. _____ cfm

11. Is this what the manufacturer rates the equipment? _____

# Oil Heating Worksheet

Introduction: When a customer complains about not enough heat from an oil-fired unit, the service technician should, as a first step, determine if the oil burner is delivering the amount of oil it was designed to deliver. Use the following procedures and the test instrument manufacturers' instructions to determine the combustion efficiency of oil burning equipment.

Tools Needed: tool kit, flue gas temperature thermometer, draft gauge, smoke tester, and carbon dioxide analyzer with the appropriate combustion efficiency chart or slide rule to use in combination with the various test results to determine the combustion efficiency.

Procedures:

1. Visually check the entire unit, and ensure all components are in proper working condition. Be sure the flue gas passages and the venting system are clear of all obstructions.

2. Is this a conversion burner? _____

3. Start the oil burner, and allow it to operate for about 15 minutes.

4. Check the flue gas temperature and record. _____ °F

5. Check the $CO_2$ content of the flue gases and record. _____ %

6. The operating combustion efficiency is _____ %

7. Conduct the smoke test and record. _____ spot

8. Measure the over-fire draft. _____ inches w.c.

9. Make any adjustments or repairs required to increase the combustion efficiency of the unit. Repeat Steps 3 through 8.

10. Measure the unit cfm and record. _____ cfm

11. Is this what the manufacturer rates the equipment? _____

# Oil Heating Worksheet

Introduction: When a customer complains about not enough heat from an oil-fired unit, the service technician should, as a first step, determine if the oil burner is delivering the amount of oil it was designed to deliver. Use the following procedures and the test instrument manufacturers' instructions to determine the combustion efficiency of oil burning equipment.

Tools Needed: tool kit, flue gas temperature thermometer, draft gauge, smoke tester, and carbon dioxide analyzer with the appropriate combustion efficiency chart or slide rule to use in combination with the various test results to determine the combustion efficiency.

Procedures:

1. Visually check the entire unit, and ensure all components are in proper working condition. Be sure the flue gas passages and the venting system are clear of all obstructions.

2. Is this a conversion burner? _____

3. Start the oil burner, and allow it to operate for about 15 minutes.

4. Check the flue gas temperature and record. _____ °F

5. Check the $CO_2$ content of the flue gases and record. _____ %

6. The operating combustion efficiency is _____ %

7. Conduct the smoke test and record. _____ spot

8. Measure the over-fire draft. _____ inches w.c.

9. Make any adjustments or repairs required to increase the combustion efficiency of the unit. Repeat Steps 3 through 8.

10. Measure the unit cfm and record. _____ cfm

11. Is this what the manufacturer rates the equipment? _____

# Oil Heating Worksheet

Introduction: When a customer complains about not enough heat from an oil-fired unit, the service technician should, as a first step, determine if the oil burner is delivering the amount of oil it was designed to deliver. Use the following procedures and the test instrument manufacturers' instructions to determine the combustion efficiency of oil burning equipment.

Tools Needed: tool kit, flue gas temperature thermometer, draft gauge, smoke tester, and carbon dioxide analyzer with the appropriate combustion efficiency chart or slide rule to use in combination with the various test results to determine the combustion efficiency.

Procedures:

1. Visually check the entire unit, and ensure all components are in proper working condition. Be sure the flue gas passages and the venting system are clear of all obstructions.

2. Is this a conversion burner? _____

3. Start the oil burner, and allow it to operate for about 15 minutes.

4. Check the flue gas temperature and record. _____°F

5. Check the $CO_2$ content of the flue gases and record. _____%

6. The operating combustion efficiency is _____%

7. Conduct the smoke test and record. _____spot

8. Measure the over-fire draft. _____inches w.c.

9. Make any adjustments or repairs required to increase the combustion efficiency of the unit. Repeat Steps 3 through 8.

10. Measure the unit cfm and record. _____cfm

11. Is this what the manufacturer rates the equipment? _____

Copyright © Business News Publishing Company

# Oil Heating Worksheet

Introduction: When a customer complains about not enough heat from an oil-fired unit, the service technician should, as a first step, determine if the oil burner is delivering the amount of oil it was designed to deliver. Use the following procedures and the test instrument manufacturers' instructions to determine the combustion efficiency of oil burning equipment.

Tools Needed: tool kit, flue gas temperature thermometer, draft gauge, smoke tester, and carbon dioxide analyzer with the appropriate combustion efficiency chart or slide rule to use in combination with the various test results to determine the combustion efficiency.

Procedures:

1. Visually check the entire unit, and ensure all components are in proper working condition. Be sure the flue gas passages and the venting system are clear of all obstructions.

2. Is this a conversion burner? _____

3. Start the oil burner, and allow it to operate for about 15 minutes.

4. Check the flue gas temperature and record. _____°F

5. Check the $CO_2$ content of the flue gases and record. _____%

6. The operating combustion efficiency is _____%

7. Conduct the smoke test and record. _____spot

8. Measure the over-fire draft. _____inches w.c.

9. Make any adjustments or repairs required to increase the combustion efficiency of the unit. Repeat Steps 3 through 8.

10. Measure the unit cfm and record. _____cfm

11. Is this what the manufacturer rates the equipment? _____

# Oil Heating Worksheet

Introduction: When a customer complains about not enough heat from an oil-fired unit, the service technician should, as a first step, determine if the oil burner is delivering the amount of oil it was designed to deliver. Use the following procedures and the test instrument manufacturers' instructions to determine the combustion efficiency of oil burning equipment.

Tools Needed: tool kit, flue gas temperature thermometer, draft gauge, smoke tester, and carbon dioxide analyzer with the appropriate combustion efficiency chart or slide rule to use in combination with the various test results to determine the combustion efficiency.

Procedures:

1. Visually check the entire unit, and ensure all components are in proper working condition. Be sure the flue gas passages and the venting system are clear of all obstructions.

2. Is this a conversion burner? _____

3. Start the oil burner, and allow it to operate for about 15 minutes.

4. Check the flue gas temperature and record. _____°F

5. Check the $CO_2$ content of the flue gases and record. _____%

6. The operating combustion efficiency is _____%

7. Conduct the smoke test and record. _____spot

8. Measure the over-fire draft. _____inches w.c.

9. Make any adjustments or repairs required to increase the combustion efficiency of the unit. Repeat Steps 3 through 8.

10. Measure the unit cfm and record. _____cfm

11. Is this what the manufacturer rates the equipment? _____

Copyright © Business News Publishing Company

# Oil Heating Worksheet

Introduction: When a customer complains about not enough heat from an oil-fired unit, the service technician should, as a first step, determine if the oil burner is delivering the amount of oil it was designed to deliver. Use the following procedures and the test instrument manufacturers' instructions to determine the combustion efficiency of oil burning equipment.

Tools Needed: tool kit, flue gas temperature thermometer, draft gauge, smoke tester, and carbon dioxide analyzer with the appropriate combustion efficiency chart or slide rule to use in combination with the various test results to determine the combustion efficiency.

Procedures:

1. Visually check the entire unit, and ensure all components are in proper working condition. Be sure the flue gas passages and the venting system are clear of all obstructions.

2. Is this a conversion burner? _____

3. Start the oil burner, and allow it to operate for about 15 minutes.

4. Check the flue gas temperature and record. _____°F

5. Check the $CO_2$ content of the flue gases and record. _____%

6. The operating combustion efficiency is _____%

7. Conduct the smoke test and record. _____spot

8. Measure the over-fire draft. _____inches w.c.

9. Make any adjustments or repairs required to increase the combustion efficiency of the unit. Repeat Steps 3 through 8.

10. Measure the unit cfm and record. _____cfm

11. Is this what the manufacturer rates the equipment? _____

# Oil Heating Worksheet

Introduction: When a customer complains about not enough heat from an oil-fired unit, the service technician should, as a first step, determine if the oil burner is delivering the amount of oil it was designed to deliver. Use the following procedures and the test instrument manufacturers' instructions to determine the combustion efficiency of oil burning equipment.

Tools Needed: tool kit, flue gas temperature thermometer, draft gauge, smoke tester, and carbon dioxide analyzer with the appropriate combustion efficiency chart or slide rule to use in combination with the various test results to determine the combustion efficiency.

Procedures:

1. Visually check the entire unit, and ensure all components are in proper working condition. Be sure the flue gas passages and the venting system are clear of all obstructions.

2. Is this a conversion burner? _____

3. Start the oil burner, and allow it to operate for about 15 minutes.

4. Check the flue gas temperature and record. _____°F

5. Check the $CO_2$ content of the flue gases and record. _____%

6. The operating combustion efficiency is _____%

7. Conduct the smoke test and record. _____spot

8. Measure the over-fire draft. _____inches w.c.

9. Make any adjustments or repairs required to increase the combustion efficiency of the unit. Repeat Steps 3 through 8.

10. Measure the unit cfm and record. _____cfm

11. Is this what the manufacturer rates the equipment? _____

# Oil Heating Worksheet

Introduction: When a customer complains about not enough heat from an oil-fired unit, the service technician should, as a first step, determine if the oil burner is delivering the amount of oil it was designed to deliver. Use the following procedures and the test instrument manufacturers' instructions to determine the combustion efficiency of oil burning equipment.

Tools Needed: tool kit, flue gas temperature thermometer, draft gauge, smoke tester, and carbon dioxide analyzer with the appropriate combustion efficiency chart or slide rule to use in combination with the various test results to determine the combustion efficiency.

Procedures:

1. Visually check the entire unit, and ensure all components are in proper working condition. Be sure the flue gas passages and the venting system are clear of all obstructions.

2. Is this a conversion burner? _____

3. Start the oil burner, and allow it to operate for about 15 minutes.

4. Check the flue gas temperature and record. _____°F

5. Check the $CO_2$ content of the flue gases and record. _____%

6. The operating combustion efficiency is _____%

7. Conduct the smoke test and record. _____spot

8. Measure the over-fire draft. _____inches w.c.

9. Make any adjustments or repairs required to increase the combustion efficiency of the unit. Repeat Steps 3 through 8.

10. Measure the unit cfm and record. _____cfm

11. Is this what the manufacturer rates the equipment? _____

Copyright © Business News Publishing Company

# Oil Heating Worksheet

Introduction: When a customer complains about not enough heat from an oil-fired unit, the service technician should, as a first step, determine if the oil burner is delivering the amount of oil it was designed to deliver. Use the following procedures and the test instrument manufacturers' instructions to determine the combustion efficiency of oil burning equipment.

Tools Needed: tool kit, flue gas temperature thermometer, draft gauge, smoke tester, and carbon dioxide analyzer with the appropriate combustion efficiency chart or slide rule to use in combination with the various test results to determine the combustion efficiency.

Procedures:

1. Visually check the entire unit, and ensure all components are in proper working condition. Be sure the flue gas passages and the venting system are clear of all obstructions.

2. Is this a conversion burner? _____

3. Start the oil burner, and allow it to operate for about 15 minutes.

4. Check the flue gas temperature and record. _____°F

5. Check the $CO_2$ content of the flue gases and record. _____%

6. The operating combustion efficiency is _____%

7. Conduct the smoke test and record. _____spot

8. Measure the over-fire draft. _____inches w.c.

9. Make any adjustments or repairs required to increase the combustion efficiency of the unit. Repeat Steps 3 through 8.

10. Measure the unit cfm and record. _____cfm

11. Is this what the manufacturer rates the equipment? _____

Copyright © Business News Publishing Company

# Air Conditioning and Heat Pump (Cooling Mode) Worksheet

Introduction: Use the following procedures and the test instrument manufacturers' instructions to determine the capacity of a cooling system.

Tools Needed: wet bulb thermometer, dry bulb thermometer, total heat content of air (Table 5-2) or psychrometric chart, tool kit, and velometer.

Procedures:

1. Set the thermostat to demand cooling.

2. Allow the unit to operate for about 15 minutes to allow the pressures and temperatures to stabilize.

3. Take the following temperature readings:

    **Indoor coil:**

    - Inlet air temperature (db) _____ °F
    - Inlet air temperature (wb) _____ °F
    - Outlet air temperature (db) _____ °F
    - Outlet air temperature (wb) _____ °F

    **Outdoor coil:**

    - Inlet air temperature (db) _____ °F
    - Outlet air temperature (db) _____ °F

4. Subtract the outlet air temperature (db) from the inlet air temperature (db) on the indoor coil and record. _____ °F

5. Subtract the inlet air temperature (db) from the outlet air temperature (db) on the outdoor coil and record. _____ °F

6. Using a psychrometric chart or Table 5-2, determine the total heat content (enthalpy) of the inlet air and record. _____ Btu/lb of dry air

7. Using a psychrometric chart or Table 5-2, determine the total heat content (enthalpy) of the outlet air and record. _____ Btu/lb of dry air

8. Determine the total heat (enthalpy) difference and record. Use the following formula:

    $\Delta H = \text{Inlet } H_t - \text{Outlet } H_t$

    $\Delta H =$ _____ Btu/lb of dry air

9. Measure the evaporator free area in square feet and record. _____ ft$^2$

10. Measure the air velocity through the indoor coil and record. _____

Copyright © Business News Publishing Company

11. Determine the indoor coil cfm and record. Use the following formula:

    cfm = Area x Velocity

    cfm = _____

12. Determine the unit capacity and record. Use the following formula:

    Btuh = cfm x 4.5 x ΔH

    Btuh = _____

13. Is this what the manufacturer rates the equipment? _____

# Air Conditioning and Heat Pump (Cooling Mode) Worksheet

Introduction: Use the following procedures and the test instrument manufacturers' instructions to determine the capacity of a cooling system.

Tools Needed: wet bulb thermometer, dry bulb thermometer, total heat content of air (Table 5-2) or psychrometric chart, tool kit, and velometer.

Procedures:

1. Set the thermostat to demand cooling.

2. Allow the unit to operate for about 15 minutes to allow the pressures and temperatures to stabilize.

3. Take the following temperature readings:

    **Indoor coil:**

    - Inlet air temperature (db) _____ °F
    - Inlet air temperature (wb) _____ °F
    - Outlet air temperature (db) _____ °F
    - Outlet air temperature (wb) _____ °F

    **Outdoor coil:**

    - Inlet air temperature (db) _____ °F
    - Outlet air temperature (db) _____ °F

4. Subtract the outlet air temperature (db) from the inlet air temperature (db) on the indoor coil and record._____°F

5. Subtract the inlet air temperature (db) from the outlet air temperature (db) on the outdoor coil and record._____°F

6. Using a psychrometric chart or Table 5-2, determine the total heat content (enthalpy) of the inlet air and record._____Btu/lb of dry air

7. Using a psychrometric chart or Table 5-2, determine the total heat content (enthalpy) of the outlet air and record._____Btu/lb of dry air

8. Determine the total heat (enthalpy) difference and record. Use the following formula:

$$\Delta H = \text{Inlet } H_t - \text{Outlet } H_t$$

$\Delta H = $ _____ Btu/lb of dry air

9. Measure the evaporator free area in square feet and record._____ft²

10. Measure the air velocity through the indoor coil and record._____

Copyright © Business News Publishing Company

11. Determine the indoor coil cfm and record. Use the following formula:

cfm = Area x Velocity

cfm = _____

12. Determine the unit capacity and record. Use the following formula:

Btuh = cfm x 4.5 x ΔH

Btuh = _____

13. Is this what the manufacturer rates the equipment? _____

# Air Conditioning and Heat Pump (Cooling Mode) Worksheet

Introduction: Use the following procedures and the test instrument manufacturers' instructions to determine the capacity of a cooling system.

Tools Needed: wet bulb thermometer, dry bulb thermometer, total heat content of air (Table 5-2) or psychrometric chart, tool kit, and velometer.

Procedures:

1. Set the thermostat to demand cooling.

2. Allow the unit to operate for about 15 minutes to allow the pressures and temperatures to stabilize.

3. Take the following temperature readings:

   **Indoor coil:**

   - Inlet air temperature (db) _____°F

   - Inlet air temperature (wb) _____°F

   - Outlet air temperature (db) _____°F

   - Outlet air temperature (wb) _____°F

   **Outdoor coil:**

   - Inlet air temperature (db) _____°F

   - Outlet air temperature (db) _____°F

4. Subtract the outlet air temperature (db) from the inlet air temperature (db) on the indoor coil and record._____°F

5. Subtract the inlet air temperature (db) from the outlet air temperature (db) on the outdoor coil and record._____°F

6. Using a psychrometric chart or Table 5-2, determine the total heat content (enthalpy) of the inlet air and record._____Btu/lb of dry air

7. Using a psychrometric chart or Table 5-2, determine the total heat content (enthalpy) of the outlet air and record._____Btu/lb of dry air

8. Determine the total heat (enthalpy) difference and record. Use the following formula:

   $$\Delta H = \text{Inlet } H_t - \text{Outlet } H_t$$

   $\Delta H = $ _____ Btu/lb of dry air

9. Measure the evaporator free area in square feet and record._____ft$^2$

10. Measure the air velocity through the indoor coil and record._____

Copyright © Business News Publishing Company

11. Determine the indoor coil cfm and record. Use the following formula:

    cfm = Area x Velocity

    cfm = _____

12. Determine the unit capacity and record. Use the following formula:

    Btuh = cfm x 4.5 x ΔH

    Btuh = _____

13. Is this what the manufacturer rates the equipment? _____

# Air Conditioning and Heat Pump (Cooling Mode) Worksheet

Introduction: Use the following procedures and the test instrument manufacturers' instructions to determine the capacity of a cooling system.

Tools Needed: wet bulb thermometer, dry bulb thermometer, total heat content of air (Table 5-2) or psychrometric chart, tool kit, and velometer.

Procedures:

1. Set the thermostat to demand cooling.

2. Allow the unit to operate for about 15 minutes to allow the pressures and temperatures to stabilize.

3. Take the following temperature readings:

    **Indoor coil:**

    - Inlet air temperature (db) _____ °F
    - Inlet air temperature (wb) _____ °F
    - Outlet air temperature (db) _____ °F
    - Outlet air temperature (wb) _____ °F

    **Outdoor coil:**

    - Inlet air temperature (db) _____ °F
    - Outlet air temperature (db) _____ °F

4. Subtract the outlet air temperature (db) from the inlet air temperature (db) on the indoor coil and record. _____ °F

5. Subtract the inlet air temperature (db) from the outlet air temperature (db) on the outdoor coil and record. _____ °F

6. Using a psychrometric chart or Table 5-2, determine the total heat content (enthalpy) of the inlet air and record. _____ Btu/lb of dry air

7. Using a psychrometric chart or Table 5-2, determine the total heat content (enthalpy) of the outlet air and record. _____ Btu/lb of dry air

8. Determine the total heat (enthalpy) difference and record. Use the following formula:

    $$\Delta H = \text{Inlet } H_t - \text{Outlet } H_t$$

    $\Delta H =$ _____ Btu/lb of dry air

9. Measure the evaporator free area in square feet and record. _____ ft$^2$

10. Measure the air velocity through the indoor coil and record. _____

11. Determine the indoor coil cfm and record. Use the following formula:

    cfm = Area x Velocity

    cfm = _____

12. Determine the unit capacity and record. Use the following formula:

    Btuh = cfm x 4.5 x ΔH

    Btuh = _____

13. Is this what the manufacturer rates the equipment? _____

# Air Conditioning and Heat Pump (Cooling Mode) Worksheet

Introduction: Use the following procedures and the test instrument manufacturers' instructions to determine the capacity of a cooling system.

Tools Needed: wet bulb thermometer, dry bulb thermometer, total heat content of air (Table 5-2) or psychrometric chart, tool kit, and velometer.

Procedures:

1. Set the thermostat to demand cooling.

2. Allow the unit to operate for about 15 minutes to allow the pressures and temperatures to stabilize.

3. Take the following temperature readings:

    **Indoor coil:**

    - Inlet air temperature (db) _____°F
    - Inlet air temperature (wb) _____°F
    - Outlet air temperature (db) _____°F
    - Outlet air temperature (wb) _____°F

    **Outdoor coil:**

    - Inlet air temperature (db) _____°F
    - Outlet air temperature (db) _____°F

4. Subtract the outlet air temperature (db) from the inlet air temperature (db) on the indoor coil and record._____°F

5. Subtract the inlet air temperature (db) from the outlet air temperature (db) on the outdoor coil and record._____°F

6. Using a psychrometric chart or Table 5-2, determine the total heat content (enthalpy) of the inlet air and record._____Btu/lb of dry air

7. Using a psychrometric chart or Table 5-2, determine the total heat content (enthalpy) of the outlet air and record._____Btu/lb of dry air

8. Determine the total heat (enthalpy) difference and record. Use the following formula:

    $$\Delta H = \text{Inlet } H_t - \text{Outlet } H_t$$

    $\Delta H = $ _____ Btu/lb of dry air

9. Measure the evaporator free area in square feet and record._____ft²

10. Measure the air velocity through the indoor coil and record._____

Copyright © Business News Publishing Company

11. Determine the indoor coil cfm and record. Use the following formula:

$$cfm = Area \times Velocity$$

cfm = _____

12. Determine the unit capacity and record. Use the following formula:

$$Btuh = cfm \times 4.5 \times \Delta H$$

Btuh = _____

13. Is this what the manufacturer rates the equipment? _____

# Air Conditioning and Heat Pump (Cooling Mode) Worksheet

Introduction: Use the following procedures and the test instrument manufacturers' instructions to determine the capacity of a cooling system.

Tools Needed: wet bulb thermometer, dry bulb thermometer, total heat content of air (Table 5-2) or psychrometric chart, tool kit, and velometer.

Procedures:

1. Set the thermostat to demand cooling.

2. Allow the unit to operate for about 15 minutes to allow the pressures and temperatures to stabilize.

3. Take the following temperature readings:

    **Indoor coil:**

    - Inlet air temperature (db) _____°F
    - Inlet air temperature (wb) _____°F
    - Outlet air temperature (db) _____°F
    - Outlet air temperature (wb) _____°F

    **Outdoor coil:**

    - Inlet air temperature (db) _____°F
    - Outlet air temperature (db) _____°F

4. Subtract the outlet air temperature (db) from the inlet air temperature (db) on the indoor coil and record._____°F

5. Subtract the inlet air temperature (db) from the outlet air temperature (db) on the outdoor coil and record._____°F

6. Using a psychrometric chart or Table 5-2, determine the total heat content (enthalpy) of the inlet air and record._____Btu/lb of dry air

7. Using a psychrometric chart or Table 5-2, determine the total heat content (enthalpy) of the outlet air and record._____Btu/lb of dry air

8. Determine the total heat (enthalpy) difference and record. Use the following formula:

    $$\Delta H = \text{Inlet } H_t - \text{Outlet } H_t$$

    $\Delta H =$ _____ Btu/lb of dry air

9. Measure the evaporator free area in square feet and record._____ft²

10. Measure the air velocity through the indoor coil and record._____

Copyright © Business News Publishing Company

11. Determine the indoor coil cfm and record. Use the following formula:

cfm = Area x Velocity

cfm = _____

12. Determine the unit capacity and record. Use the following formula:

Btuh = cfm x 4.5 x $\Delta$H

Btuh = _____

13. Is this what the manufacturer rates the equipment? _____

# Air Conditioning and Heat Pump (Cooling Mode) Worksheet

Introduction: Use the following procedures and the test instrument manufacturers' instructions to determine the capacity of a cooling system.

Tools Needed: wet bulb thermometer, dry bulb thermometer, total heat content of air (Table 5-2) or psychrometric chart, tool kit, and velometer.

Procedures:

1. Set the thermostat to demand cooling.

2. Allow the unit to operate for about 15 minutes to allow the pressures and temperatures to stabilize.

3. Take the following temperature readings:

    **Indoor coil:**

    - Inlet air temperature (db) _____°F

    - Inlet air temperature (wb) _____°F

    - Outlet air temperature (db) _____°F

    - Outlet air temperature (wb) _____°F

    **Outdoor coil:**

    - Inlet air temperature (db) _____°F

    - Outlet air temperature (db) _____°F

4. Subtract the outlet air temperature (db) from the inlet air temperature (db) on the indoor coil and record._____°F

5. Subtract the inlet air temperature (db) from the outlet air temperature (db) on the outdoor coil and record._____°F

6. Using a psychrometric chart or Table 5-2, determine the total heat content (enthalpy) of the inlet air and record._____Btu/lb of dry air

7. Using a psychrometric chart or Table 5-2, determine the total heat content (enthalpy) of the outlet air and record._____Btu/lb of dry air

8. Determine the total heat (enthalpy) difference and record. Use the following formula:

    $$\Delta H = \text{Inlet } H_t - \text{Outlet } H_t$$

    $\Delta H = $ _____ Btu/lb of dry air

9. Measure the evaporator free area in square feet and record._____ft$^2$

10. Measure the air velocity through the indoor coil and record._____

Copyright © Business News Publishing Company

11. Determine the indoor coil cfm and record. Use the following formula:

$$cfm = Area \times Velocity$$

cfm = _____

12. Determine the unit capacity and record. Use the following formula:

$$Btuh = cfm \times 4.5 \times \Delta H$$

Btuh = _____

13. Is this what the manufacturer rates the equipment? _____

# Air Conditioning and Heat Pump (Cooling Mode) Worksheet

Introduction: Use the following procedures and the test instrument manufacturers' instructions to determine the capacity of a cooling system.

Tools Needed: wet bulb thermometer, dry bulb thermometer, total heat content of air (Table 5-2) or psychrometric chart, tool kit, and velometer.

Procedures:

1. Set the thermostat to demand cooling.

2. Allow the unit to operate for about 15 minutes to allow the pressures and temperatures to stabilize.

3. Take the following temperature readings:

    **Indoor coil:**

    - Inlet air temperature (db) _____ °F
    - Inlet air temperature (wb) _____ °F
    - Outlet air temperature (db) _____ °F
    - Outlet air temperature (wb) _____ °F

    **Outdoor coil:**

    - Inlet air temperature (db) _____ °F
    - Outlet air temperature (db) _____ °F

4. Subtract the outlet air temperature (db) from the inlet air temperature (db) on the indoor coil and record. _____ °F

5. Subtract the inlet air temperature (db) from the outlet air temperature (db) on the outdoor coil and record. _____ °F

6. Using a psychrometric chart or Table 5-2, determine the total heat content (enthalpy) of the inlet air and record. _____ Btu/lb of dry air

7. Using a psychrometric chart or Table 5-2, determine the total heat content (enthalpy) of the outlet air and record. _____ Btu/lb of dry air

8. Determine the total heat (enthalpy) difference and record. Use the following formula:

    $$\Delta H = \text{Inlet } H_t - \text{Outlet } H_t$$

    $\Delta H = $ _____ Btu/lb of dry air

9. Measure the evaporator free area in square feet and record. _____ ft$^2$

10. Measure the air velocity through the indoor coil and record. _____

Copyright © Business News Publishing Company

11. Determine the indoor coil cfm and record. Use the following formula:

    cfm = Area x Velocity

    cfm = _____

12. Determine the unit capacity and record. Use the following formula:

    Btuh = cfm x 4.5 x $\Delta$H

    Btuh = _____

13. Is this what the manufacturer rates the equipment? _____

# Air Conditioning and Heat Pump (Cooling Mode) Worksheet

Introduction: Use the following procedures and the test instrument manufacturers' instructions to determine the capacity of a cooling system.

Tools Needed: wet bulb thermometer, dry bulb thermometer, total heat content of air (Table 5-2) or psychrometric chart, tool kit, and velometer.

Procedures:

1. Set the thermostat to demand cooling.

2. Allow the unit to operate for about 15 minutes to allow the pressures and temperatures to stabilize.

3. Take the following temperature readings:

    **Indoor coil:**

    - Inlet air temperature (db) _____ °F
    - Inlet air temperature (wb) _____ °F
    - Outlet air temperature (db) _____ °F
    - Outlet air temperature (wb) _____ °F

    **Outdoor coil:**

    - Inlet air temperature (db) _____ °F
    - Outlet air temperature (db) _____ °F

4. Subtract the outlet air temperature (db) from the inlet air temperature (db) on the indoor coil and record. _____ °F

5. Subtract the inlet air temperature (db) from the outlet air temperature (db) on the outdoor coil and record. _____ °F

6. Using a psychrometric chart or Table 5-2, determine the total heat content (enthalpy) of the inlet air and record. _____ Btu/lb of dry air

7. Using a psychrometric chart or Table 5-2, determine the total heat content (enthalpy) of the outlet air and record. _____ Btu/lb of dry air

8. Determine the total heat (enthalpy) difference and record. Use the following formula:

    $$\Delta H = \text{Inlet } H_t - \text{Outlet } H_t$$

    $\Delta H = $ _____ Btu/lb of dry air

9. Measure the evaporator free area in square feet and record. _____ ft$^2$

10. Measure the air velocity through the indoor coil and record. _____

Copyright © Business News Publishing Company

11. Determine the indoor coil cfm and record. Use the following formula:

$$cfm = Area \times Velocity$$

cfm = _____

12. Determine the unit capacity and record. Use the following formula:

$$Btuh = cfm \times 4.5 \times \Delta H$$

Btuh = _____

13. Is this what the manufacturer rates the equipment? _____

# Air Conditioning and Heat Pump (Cooling Mode) Worksheet

Introduction: Use the following procedures and the test instrument manufacturers' instructions to determine the capacity of a cooling system.

Tools Needed: wet bulb thermometer, dry bulb thermometer, total heat content of air (Table 5-2) or psychrometric chart, tool kit, and velometer.

Procedures:

1. Set the thermostat to demand cooling.

2. Allow the unit to operate for about 15 minutes to allow the pressures and temperatures to stabilize.

3. Take the following temperature readings:

   **Indoor coil:**

   - Inlet air temperature (db) _____ °F

   - Inlet air temperature (wb) _____ °F

   - Outlet air temperature (db) _____ °F

   - Outlet air temperature (wb) _____ °F

   **Outdoor coil:**

   - Inlet air temperature (db) _____ °F

   - Outlet air temperature (db) _____ °F

4. Subtract the outlet air temperature (db) from the inlet air temperature (db) on the indoor coil and record._____°F

5. Subtract the inlet air temperature (db) from the outlet air temperature (db) on the outdoor coil and record._____°F

6. Using a psychrometric chart or Table 5-2, determine the total heat content (enthalpy) of the inlet air and record._____Btu/lb of dry air

7. Using a psychrometric chart or Table 5-2, determine the total heat content (enthalpy) of the outlet air and record._____Btu/lb of dry air

8. Determine the total heat (enthalpy) difference and record. Use the following formula:

    $\Delta H = $ Inlet $H_t$ - Outlet $H_t$

    $\Delta H = $ _____ Btu/lb of dry air

9. Measure the evaporator free area in square feet and record._____ft²

10. Measure the air velocity through the indoor coil and record._____

Copyright © Business News Publishing Company

11. Determine the indoor coil cfm and record. Use the following formula:

    cfm = Area x Velocity

    cfm = _____

12. Determine the unit capacity and record. Use the following formula:

    Btuh = cfm x 4.5 x ΔH

    Btuh = _____

13. Is this what the manufacturer rates the equipment? _____

# Air Conditioning and Heat Pump (Cooling Mode) Worksheet

Introduction: Use the following procedures and the test instrument manufacturers' instructions to determine the capacity of a cooling system.

Tools Needed: wet bulb thermometer, dry bulb thermometer, total heat content of air (Table 5-2) or psychrometric chart, tool kit, and velometer.

Procedures:

1. Set the thermostat to demand cooling.

2. Allow the unit to operate for about 15 minutes to allow the pressures and temperatures to stabilize.

3. Take the following temperature readings:

    **Indoor coil:**

    - Inlet air temperature (db) _____°F

    - Inlet air temperature (wb) _____°F

    - Outlet air temperature (db) _____°F

    - Outlet air temperature (wb) _____°F

    **Outdoor coil:**

    - Inlet air temperature (db) _____°F

    - Outlet air temperature (db) _____°F

4. Subtract the outlet air temperature (db) from the inlet air temperature (db) on the indoor coil and record._____°F

5. Subtract the inlet air temperature (db) from the outlet air temperature (db) on the outdoor coil and record._____°F

6. Using a psychrometric chart or Table 5-2, determine the total heat content (enthalpy) of the inlet air and record._____Btu/lb of dry air

7. Using a psychrometric chart or Table 5-2, determine the total heat content (enthalpy) of the outlet air and record._____Btu/lb of dry air

8. Determine the total heat (enthalpy) difference and record. Use the following formula:

    $$\Delta H = \text{Inlet } H_t - \text{Outlet } H_t$$

    $\Delta H = $ _____Btu/lb of dry air

9. Measure the evaporator free area in square feet and record._____ft$^2$

10. Measure the air velocity through the indoor coil and record._____

Copyright © Business News Publishing Company

11. Determine the indoor coil cfm and record. Use the following formula:

    cfm = Area x Velocity

    cfm = _____

12. Determine the unit capacity and record. Use the following formula:

    Btuh = cfm x 4.5 x ΔH

    Btuh = _____

13. Is this what the manufacturer rates the equipment? _____

# Air Conditioning and Heat Pump (Cooling Mode) Worksheet

Introduction: Use the following procedures and the test instrument manufacturers' instructions to determine the capacity of a cooling system.

Tools Needed: wet bulb thermometer, dry bulb thermometer, total heat content of air (Table 5-2) or psychrometric chart, tool kit, and velometer.

Procedures:

1. Set the thermostat to demand cooling.

2. Allow the unit to operate for about 15 minutes to allow the pressures and temperatures to stabilize.

3. Take the following temperature readings:

    **Indoor coil:**

    - Inlet air temperature (db) _____°F

    - Inlet air temperature (wb) _____°F

    - Outlet air temperature (db) _____°F

    - Outlet air temperature (wb) _____°F

    **Outdoor coil:**

    - Inlet air temperature (db) _____°F

    - Outlet air temperature (db) _____°F

4. Subtract the outlet air temperature (db) from the inlet air temperature (db) on the indoor coil and record._____°F

5. Subtract the inlet air temperature (db) from the outlet air temperature (db) on the outdoor coil and record._____°F

6. Using a psychrometric chart or Table 5-2, determine the total heat content (enthalpy) of the inlet air and record._____Btu/lb of dry air

7. Using a psychrometric chart or Table 5-2, determine the total heat content (enthalpy) of the outlet air and record._____Btu/lb of dry air

8. Determine the total heat (enthalpy) difference and record. Use the following formula:

    $\Delta H = $ Inlet $H_t$ - Outlet $H_t$

    $\Delta H = $ _____ Btu/lb of dry air

9. Measure the evaporator free area in square feet and record._____ft$^2$

10. Measure the air velocity through the indoor coil and record._____

Copyright © Business News Publishing Company

11. Determine the indoor coil cfm and record. Use the following formula:

    cfm = Area x Velocity

    cfm = _____

12. Determine the unit capacity and record. Use the following formula:

    Btuh = cfm x 4.5 x ΔH

    Btuh = _____

13. Is this what the manufacturer rates the equipment? _____

# Air Conditioning and Heat Pump (Cooling Mode) Worksheet

Introduction: Use the following procedures and the test instrument manufacturers' instructions to determine the capacity of a cooling system.

Tools Needed: wet bulb thermometer, dry bulb thermometer, total heat content of air (Table 5-2) or psychrometric chart, tool kit, and velometer.

Procedures:

1. Set the thermostat to demand cooling.

2. Allow the unit to operate for about 15 minutes to allow the pressures and temperatures to stabilize.

3. Take the following temperature readings:

    **Indoor coil:**

    - Inlet air temperature (db) _____ °F

    - Inlet air temperature (wb) _____ °F

    - Outlet air temperature (db) _____ °F

    - Outlet air temperature (wb) _____ °F

    **Outdoor coil:**

    - Inlet air temperature (db) _____ °F

    - Outlet air temperature (db) _____ °F

4. Subtract the outlet air temperature (db) from the inlet air temperature (db) on the indoor coil and record. _____ °F

5. Subtract the inlet air temperature (db) from the outlet air temperature (db) on the outdoor coil and record. _____ °F

6. Using a psychrometric chart or Table 5-2, determine the total heat content (enthalpy) of the inlet air and record. _____ Btu/lb of dry air

7. Using a psychrometric chart or Table 5-2, determine the total heat content (enthalpy) of the outlet air and record. _____ Btu/lb of dry air

8. Determine the total heat (enthalpy) difference and record. Use the following formula:

    $$\Delta H = \text{Inlet } H_t - \text{Outlet } H_t$$

    $\Delta H =$ _____ Btu/lb of dry air

9. Measure the evaporator free area in square feet and record. _____ ft$^2$

10. Measure the air velocity through the indoor coil and record. _____

Copyright © Business News Publishing Company

11. Determine the indoor coil cfm and record. Use the following formula:

    cfm = Area x Velocity

    cfm = _____

12. Determine the unit capacity and record. Use the following formula:

    Btuh = cfm x 4.5 x ΔH

    Btuh = _____

13. Is this what the manufacturer rates the equipment? _____

# Air Conditioning and Heat Pump (Cooling Mode) Worksheet

Introduction: Use the following procedures and the test instrument manufacturers' instructions to determine the capacity of a cooling system.

Tools Needed: wet bulb thermometer, dry bulb thermometer, total heat content of air (Table 5-2) or psychrometric chart, tool kit, and velometer.

Procedures:

1. Set the thermostat to demand cooling.

2. Allow the unit to operate for about 15 minutes to allow the pressures and temperatures to stabilize.

3. Take the following temperature readings:

    **Indoor coil:**

    - Inlet air temperature (db) _____°F
    - Inlet air temperature (wb) _____°F
    - Outlet air temperature (db) _____°F
    - Outlet air temperature (wb) _____°F

    **Outdoor coil:**

    - Inlet air temperature (db) _____°F
    - Outlet air temperature (db) _____°F

4. Subtract the outlet air temperature (db) from the inlet air temperature (db) on the indoor coil and record._____°F

5. Subtract the inlet air temperature (db) from the outlet air temperature (db) on the outdoor coil and record._____°F

6. Using a psychrometric chart or Table 5-2, determine the total heat content (enthalpy) of the inlet air and record._____Btu/lb of dry air

7. Using a psychrometric chart or Table 5-2, determine the total heat content (enthalpy) of the outlet air and record._____Btu/lb of dry air

8. Determine the total heat (enthalpy) difference and record. Use the following formula:

    $$\Delta H = \text{Inlet } H_t - \text{Outlet } H_t$$

    $\Delta H = $ _____ Btu/lb of dry air

9. Measure the evaporator free area in square feet and record._____ft$^2$

10. Measure the air velocity through the indoor coil and record._____

Copyright © Business News Publishing Company

11. Determine the indoor coil cfm and record. Use the following formula:

$$\text{cfm} = \text{Area} \times \text{Velocity}$$

cfm = _____

12. Determine the unit capacity and record. Use the following formula:

$$\text{Btuh} = \text{cfm} \times 4.5 \times \Delta H$$

Btuh = _____

13. Is this what the manufacturer rates the equipment? _____

# Air Conditioning and Heat Pump (Cooling Mode) Worksheet

Introduction: Use the following procedures and the test instrument manufacturers' instructions to determine the capacity of a cooling system.

Tools Needed: wet bulb thermometer, dry bulb thermometer, total heat content of air (Table 5-2) or psychrometric chart, tool kit, and velometer.

Procedures:

1. Set the thermostat to demand cooling.

2. Allow the unit to operate for about 15 minutes to allow the pressures and temperatures to stabilize.

3. Take the following temperature readings:

    **Indoor coil:**

    - Inlet air temperature (db) _____ °F
    - Inlet air temperature (wb) _____ °F
    - Outlet air temperature (db) _____ °F
    - Outlet air temperature (wb) _____ °F

    **Outdoor coil:**

    - Inlet air temperature (db) _____ °F
    - Outlet air temperature (db) _____ °F

4. Subtract the outlet air temperature (db) from the inlet air temperature (db) on the indoor coil and record. _____ °F

5. Subtract the inlet air temperature (db) from the outlet air temperature (db) on the outdoor coil and record. _____ °F

6. Using a psychrometric chart or Table 5-2, determine the total heat content (enthalpy) of the inlet air and record. _____ Btu/lb of dry air

7. Using a psychrometric chart or Table 5-2, determine the total heat content (enthalpy) of the outlet air and record. _____ Btu/lb of dry air

8. Determine the total heat (enthalpy) difference and record. Use the following formula:

$$\Delta H = \text{Inlet } H_t - \text{Outlet } H_t$$

$\Delta H = $ _____ Btu/lb of dry air

9. Measure the evaporator free area in square feet and record. _____ ft$^2$

10. Measure the air velocity through the indoor coil and record. _____

Copyright © Business News Publishing Company

11. Determine the indoor coil cfm and record. Use the following formula:

    cfm = Area x Velocity

    cfm = _____

12. Determine the unit capacity and record. Use the following formula:

    Btuh = cfm x 4.5 x $\Delta H$

    Btuh = _____

13. Is this what the manufacturer rates the equipment? _____

# Heat Pump (Heating Mode) Worksheet

Introduction: The service technician is often called upon to perform a capacity check of a heat pump system. This request may come in a variety of forms, such as a complaint of a high electric bill or not enough heat. Determining the capacity of a heat pump in the heating mode is not very difficult, and every service technician should perform this test whenever there is a suspected problem with the unit.

Tools Needed: dry bulb thermometer, velometer, gauge manifold, voltmeter, wattmeter, and tool kit.

Procedures:

1. Set the room thermostat to the heating or automatic position.

2. Set the temperature lever to demand heating.

3. Turn off all electricity to the auxiliary heat strips. Only the heat pump and the indoor fan are to be operating during this test.

4. Measure the temperature rise of the air through the indoor unit. Use the following formula:

    $$\Delta T = \text{Leaving air temperature} - \text{Entering air temperature}$$

    a. Use the same thermometer for measuring both the return and supply air temperatures.

    b. Do not measure the temperature "in view" of the indoor coil. True air temperature cannot be measured in areas affected by radiant heat.

    c. Make the temperature measurements within 6 ft of the indoor unit. Measurements taken at the supply and return air grilles are not accurate enough.

    d. Use the average temperature when more than one duct is connected to the plenum. Use the following formula:

    $$\text{Average} = \frac{\text{Total of readings}}{\text{Number of readings}}$$

    e. Make sure the air temperature is stable before taking these measurements.

    f. Take these measurements downstream from any mixed air source.

    g. Record the temperature difference of the return and supply air as $\Delta T$.

    $$\Delta T = \underline{\qquad} - \underline{\qquad}$$

Copyright © Business News Publishing Company

5. Determine the Btu output. Use the following formula:

$$Btu = cfm \times 1.08 \times \Delta T$$

where: $cfm = \dfrac{Btuh}{1.08 \times \Delta T}$

1.08 = specific heat of air constant

$\Delta T$ = supply air temperature minus return air temperature

6. Is this what the manufacturer rates the equipment? _____

# Heat Pump (Heating Mode) Worksheet

Introduction: The service technician is often called upon to perform a capacity check of a heat pump system. This request may come in a variety of forms, such as a complaint of a high electric bill or not enough heat. Determining the capacity of a heat pump in the heating mode is not very difficult, and every service technician should perform this test whenever there is a suspected problem with the unit.

Tools Needed: dry bulb thermometer, velometer, gauge manifold, voltmeter, wattmeter, and tool kit.

Procedures:

1. Set the room thermostat to the heating or automatic position.

2. Set the temperature lever to demand heating.

3. Turn off all electricity to the auxiliary heat strips. Only the heat pump and the indoor fan are to be operating during this test.

4. Measure the temperature rise of the air through the indoor unit. Use the following formula:

    $\Delta T$ = Leaving air temperature - Entering air temperature

    a. Use the same thermometer for measuring both the return and supply air temperatures.

    b. Do not measure the temperature "in view" of the indoor coil. True air temperature cannot be measured in areas affected by radiant heat.

    c. Make the temperature measurements within 6 ft of the indoor unit. Measurements taken at the supply and return air grilles are not accurate enough.

    d. Use the average temperature when more than one duct is connected to the plenum. Use the following formula:

    $$\text{Average} = \frac{\text{Total of readings}}{\text{Number of readings}}$$

    e. Make sure the air temperature is stable before taking these measurements.

    f. Take these measurements downstream from any mixed air source.

    g. Record the temperature difference of the return and supply air as $\Delta T$.

    $\Delta T$ = _____ - _____

Copyright © Business News Publishing Company

5. Determine the Btu output. Use the following formula:

$$\text{Btu} = \text{cfm} \times 1.08 \times \Delta T$$

where: $\text{cfm} = \dfrac{\text{Btuh}}{1.08 \times \Delta T}$

1.08 = specific heat of air constant

$\Delta T$ = supply air temperature minus return air temperature

6. Is this what the manufacturer rates the equipment? _____

# Heat Pump (Heating Mode) Worksheet

Introduction: The service technician is often called upon to perform a capacity check of a heat pump system. This request may come in a variety of forms, such as a complaint of a high electric bill or not enough heat. Determining the capacity of a heat pump in the heating mode is not very difficult, and every service technician should perform this test whenever there is a suspected problem with the unit.

Tools Needed: dry bulb thermometer, velometer, gauge manifold, voltmeter, wattmeter, and tool kit.

Procedures:

1. Set the room thermostat to the heating or automatic position.

2. Set the temperature lever to demand heating.

3. Turn off all electricity to the auxiliary heat strips. Only the heat pump and the indoor fan are to be operating during this test.

4. Measure the temperature rise of the air through the indoor unit. Use the following formula:

    $\Delta T$ = Leaving air temperature - Entering air temperature

    a. Use the same thermometer for measuring both the return and supply air temperatures.

    b. Do not measure the temperature "in view" of the indoor coil. True air temperature cannot be measured in areas affected by radiant heat.

    c. Make the temperature measurements within 6 ft of the indoor unit. Measurements taken at the supply and return air grilles are not accurate enough.

    d. Use the average temperature when more than one duct is connected to the plenum. Use the following formula:

    $$\text{Average} = \frac{\text{Total of readings}}{\text{Number of readings}}$$

    e. Make sure the air temperature is stable before taking these measurements.

    f. Take these measurements downstream from any mixed air source.

    g. Record the temperature difference of the return and supply air as $\Delta T$.

    $\Delta T$ = _____ - _____

5. Determine the Btu output. Use the following formula:

$$Btu = cfm \times 1.08 \times \Delta T$$

where: $cfm = \dfrac{Btuh}{1.08 \times \Delta T}$

1.08 = specific heat of air constant

$\Delta T$ = supply air temperature minus return air temperature

6. Is this what the manufacturer rates the equipment? _____

# Heat Pump (Heating Mode) Worksheet

Introduction: The service technician is often called upon to perform a capacity check of a heat pump system. This request may come in a variety of forms, such as a complaint of a high electric bill or not enough heat. Determining the capacity of a heat pump in the heating mode is not very difficult, and every service technician should perform this test whenever there is a suspected problem with the unit.

Tools Needed: dry bulb thermometer, velometer, gauge manifold, voltmeter, wattmeter, and tool kit.

Procedures:

1. Set the room thermostat to the heating or automatic position.

2. Set the temperature lever to demand heating.

3. Turn off all electricity to the auxiliary heat strips. Only the heat pump and the indoor fan are to be operating during this test.

4. Measure the temperature rise of the air through the indoor unit. Use the following formula:

    $\Delta T$ = Leaving air temperature - Entering air temperature

    a. Use the same thermometer for measuring both the return and supply air temperatures.

    b. Do not measure the temperature "in view" of the indoor coil. True air temperature cannot be measured in areas affected by radiant heat.

    c. Make the temperature measurements within 6 ft of the indoor unit. Measurements taken at the supply and return air grilles are not accurate enough.

    d. Use the average temperature when more than one duct is connected to the plenum. Use the following formula:

    $$\text{Average} = \frac{\text{Total of readings}}{\text{Number of readings}}$$

    e. Make sure the air temperature is stable before taking these measurements.

    f. Take these measurements downstream from any mixed air source.

    g. Record the temperature difference of the return and supply air as $\Delta T$.

    $\Delta T$ = _____ - _____

Copyright © Business News Publishing Company

5. Determine the Btu output. Use the following formula:

$$Btu = cfm \times 1.08 \times \Delta T$$

where: $cfm = \dfrac{Btuh}{1.08 \times \Delta T}$

$1.08$ = specific heat of air constant

$\Delta T$ = supply air temperature minus return air temperature

6. Is this what the manufacturer rates the equipment? _____

# Heat Pump (Heating Mode) Worksheet

Introduction: The service technician is often called upon to perform a capacity check of a heat pump system. This request may come in a variety of forms, such as a complaint of a high electric bill or not enough heat. Determining the capacity of a heat pump in the heating mode is not very difficult, and every service technician should perform this test whenever there is a suspected problem with the unit.

Tools Needed: dry bulb thermometer, velometer, gauge manifold, voltmeter, wattmeter, and tool kit.

Procedures:

1. Set the room thermostat to the heating or automatic position.

2. Set the temperature lever to demand heating.

3. Turn off all electricity to the auxiliary heat strips. Only the heat pump and the indoor fan are to be operating during this test.

4. Measure the temperature rise of the air through the indoor unit. Use the following formula:

    $\Delta T$ = Leaving air temperature - Entering air temperature

    a. Use the same thermometer for measuring both the return and supply air temperatures.

    b. Do not measure the temperature "in view" of the indoor coil. True air temperature cannot be measured in areas affected by radiant heat.

    c. Make the temperature measurements within 6 ft of the indoor unit. Measurements taken at the supply and return air grilles are not accurate enough.

    d. Use the average temperature when more than one duct is connected to the plenum. Use the following formula:

    $$\text{Average} = \frac{\text{Total of readings}}{\text{Number of readings}}$$

    e. Make sure the air temperature is stable before taking these measurements.

    f. Take these measurements downstream from any mixed air source.

    g. Record the temperature difference of the return and supply air as $\Delta T$.

    $\Delta T$ = _____ - _____

5. Determine the Btu output. Use the following formula:

$$Btu = cfm \times 1.08 \times \Delta T$$

where: $cfm = \dfrac{Btuh}{1.08 \times \Delta T}$

1.08 = specific heat of air constant

$\Delta T$ = supply air temperature minus return air temperature

6. Is this what the manufacturer rates the equipment? _____

# Heat Pump (Heating Mode) Worksheet

Introduction: The service technician is often called upon to perform a capacity check of a heat pump system. This request may come in a variety of forms, such as a complaint of a high electric bill or not enough heat. Determining the capacity of a heat pump in the heating mode is not very difficult, and every service technician should perform this test whenever there is a suspected problem with the unit.

Tools Needed: dry bulb thermometer, velometer, gauge manifold, voltmeter, wattmeter, and tool kit.

Procedures:

1. Set the room thermostat to the heating or automatic position.

2. Set the temperature lever to demand heating.

3. Turn off all electricity to the auxiliary heat strips. Only the heat pump and the indoor fan are to be operating during this test.

4. Measure the temperature rise of the air through the indoor unit. Use the following formula:

    $\Delta T$ = Leaving air temperature - Entering air temperature

    a. Use the same thermometer for measuring both the return and supply air temperatures.

    b. Do not measure the temperature "in view" of the indoor coil. True air temperature cannot be measured in areas affected by radiant heat.

    c. Make the temperature measurements within 6 ft of the indoor unit. Measurements taken at the supply and return air grilles are not accurate enough.

    d. Use the average temperature when more than one duct is connected to the plenum. Use the following formula:

    $$\text{Average} = \frac{\text{Total of readings}}{\text{Number of readings}}$$

    e. Make sure the air temperature is stable before taking these measurements.

    f. Take these measurements downstream from any mixed air source.

    g. Record the temperature difference of the return and supply air as $\Delta T$.

    $\Delta T$ = _____ - _____

5. Determine the Btu output. Use the following formula:

$$Btu = cfm \times 1.08 \times \Delta T$$

where: $cfm = \dfrac{Btuh}{1.08 \times \Delta T}$

1.08 = specific heat of air constant

$\Delta T$ = supply air temperature minus return air temperature

6. Is this what the manufacturer rates the equipment? _____

# Heat Pump (Heating Mode) Worksheet

Introduction: The service technician is often called upon to perform a capacity check of a heat pump system. This request may come in a variety of forms, such as a complaint of a high electric bill or not enough heat. Determining the capacity of a heat pump in the heating mode is not very difficult, and every service technician should perform this test whenever there is a suspected problem with the unit.

Tools Needed: dry bulb thermometer, velometer, gauge manifold, voltmeter, wattmeter, and tool kit.

Procedures:

1. Set the room thermostat to the heating or automatic position.

2. Set the temperature lever to demand heating.

3. Turn off all electricity to the auxiliary heat strips. Only the heat pump and the indoor fan are to be operating during this test.

4. Measure the temperature rise of the air through the indoor unit. Use the following formula:

    $$\Delta T = \text{Leaving air temperature} - \text{Entering air temperature}$$

    a. Use the same thermometer for measuring both the return and supply air temperatures.

    b. Do not measure the temperature "in view" of the indoor coil. True air temperature cannot be measured in areas affected by radiant heat.

    c. Make the temperature measurements within 6 ft of the indoor unit. Measurements taken at the supply and return air grilles are not accurate enough.

    d. Use the average temperature when more than one duct is connected to the plenum. Use the following formula:

    $$\text{Average} = \frac{\text{Total of readings}}{\text{Number of readings}}$$

    e. Make sure the air temperature is stable before taking these measurements.

    f. Take these measurements downstream from any mixed air source.

    g. Record the temperature difference of the return and supply air as $\Delta T$.

    $$\Delta T = \underline{\qquad} - \underline{\qquad}$$

Copyright © Business News Publishing Company

5. Determine the Btu output. Use the following formula:

$$Btu = cfm \times 1.08 \times \Delta T$$

where: $cfm = \dfrac{Btuh}{1.08 \times \Delta T}$

1.08 = specific heat of air constant

$\Delta T$ = supply air temperature minus return air temperature

6. Is this what the manufacturer rates the equipment? _____

# Heat Pump (Heating Mode) Worksheet

Introduction: The service technician is often called upon to perform a capacity check of a heat pump system. This request may come in a variety of forms, such as a complaint of a high electric bill or not enough heat. Determining the capacity of a heat pump in the heating mode is not very difficult, and every service technician should perform this test whenever there is a suspected problem with the unit.

Tools Needed: dry bulb thermometer, velometer, gauge manifold, voltmeter, wattmeter, and tool kit.

Procedures:

1. Set the room thermostat to the heating or automatic position.

2. Set the temperature lever to demand heating.

3. Turn off all electricity to the auxiliary heat strips. Only the heat pump and the indoor fan are to be operating during this test.

4. Measure the temperature rise of the air through the indoor unit. Use the following formula:

    $$\Delta T = \text{Leaving air temperature} - \text{Entering air temperature}$$

    a. Use the same thermometer for measuring both the return and supply air temperatures.

    b. Do not measure the temperature "in view" of the indoor coil. True air temperature cannot be measured in areas affected by radiant heat.

    c. Make the temperature measurements within 6 ft of the indoor unit. Measurements taken at the supply and return air grilles are not accurate enough.

    d. Use the average temperature when more than one duct is connected to the plenum. Use the following formula:

    $$\text{Average} = \frac{\text{Total of readings}}{\text{Number of readings}}$$

    e. Make sure the air temperature is stable before taking these measurements.

    f. Take these measurements downstream from any mixed air source.

    g. Record the temperature difference of the return and supply air as $\Delta T$.

    $$\Delta T = \underline{\qquad} - \underline{\qquad}$$

Copyright © Business News Publishing Company

5. Determine the Btu output. Use the following formula:

$$Btu = cfm \times 1.08 \times \Delta T$$

where: $cfm = \dfrac{Btuh}{1.08 \times \Delta T}$

1.08 = specific heat of air constant

$\Delta T$ = supply air temperature minus return air temperature

6. Is this what the manufacturer rates the equipment? _____

# Heat Pump (Heating Mode) Worksheet

Introduction: The service technician is often called upon to perform a capacity check of a heat pump system. This request may come in a variety of forms, such as a complaint of a high electric bill or not enough heat. Determining the capacity of a heat pump in the heating mode is not very difficult, and every service technician should perform this test whenever there is a suspected problem with the unit.

Tools Needed: dry bulb thermometer, velometer, gauge manifold, voltmeter, wattmeter, and tool kit.

Procedures:

1. Set the room thermostat to the heating or automatic position.

2. Set the temperature lever to demand heating.

3. Turn off all electricity to the auxiliary heat strips. Only the heat pump and the indoor fan are to be operating during this test.

4. Measure the temperature rise of the air through the indoor unit. Use the following formula:

    $\Delta T$ = Leaving air temperature - Entering air temperature

    a. Use the same thermometer for measuring both the return and supply air temperatures.

    b. Do not measure the temperature "in view" of the indoor coil. True air temperature cannot be measured in areas affected by radiant heat.

    c. Make the temperature measurements within 6 ft of the indoor unit. Measurements taken at the supply and return air grilles are not accurate enough.

    d. Use the average temperature when more than one duct is connected to the plenum. Use the following formula:

    $$\text{Average} = \frac{\text{Total of readings}}{\text{Number of readings}}$$

    e. Make sure the air temperature is stable before taking these measurements.

    f. Take these measurements downstream from any mixed air source.

    g. Record the temperature difference of the return and supply air as $\Delta T$.

    $\Delta T$ = _____ - _____

Copyright © Business News Publishing Company

5. Determine the Btu output. Use the following formula:

$$Btu = cfm \times 1.08 \times \Delta T$$

where: $cfm = \dfrac{Btuh}{1.08 \times \Delta T}$

1.08 = specific heat of air constant

$\Delta T$ = supply air temperature minus return air temperature

6. Is this what the manufacturer rates the equipment? _____

# Heat Pump (Heating Mode) Worksheet

Introduction: The service technician is often called upon to perform a capacity check of a heat pump system. This request may come in a variety of forms, such as a complaint of a high electric bill or not enough heat. Determining the capacity of a heat pump in the heating mode is not very difficult, and every service technician should perform this test whenever there is a suspected problem with the unit.

Tools Needed: dry bulb thermometer, velometer, gauge manifold, voltmeter, wattmeter, and tool kit.

Procedures:

1. Set the room thermostat to the heating or automatic position.

2. Set the temperature lever to demand heating.

3. Turn off all electricity to the auxiliary heat strips. Only the heat pump and the indoor fan are to be operating during this test.

4. Measure the temperature rise of the air through the indoor unit. Use the following formula:

$$\Delta T = \text{Leaving air temperature} - \text{Entering air temperature}$$

   a. Use the same thermometer for measuring both the return and supply air temperatures.

   b. Do not measure the temperature "in view" of the indoor coil. True air temperature cannot be measured in areas affected by radiant heat.

   c. Make the temperature measurements within 6 ft of the indoor unit. Measurements taken at the supply and return air grilles are not accurate enough.

   d. Use the average temperature when more than one duct is connected to the plenum. Use the following formula:

$$\text{Average} = \frac{\text{Total of readings}}{\text{Number of readings}}$$

   e. Make sure the air temperature is stable before taking these measurements.

   f. Take these measurements downstream from any mixed air source.

   g. Record the temperature difference of the return and supply air as $\Delta T$.

$$\Delta T = \underline{\qquad} - \underline{\qquad}$$

Copyright © Business News Publishing Company

5. Determine the Btu output. Use the following formula:

$$Btu = cfm \times 1.08 \times \Delta T$$

where: $cfm = \dfrac{Btuh}{1.08 \times \Delta T}$

1.08 = specific heat of air constant

$\Delta T$ = supply air temperature minus return air temperature

6. Is this what the manufacturer rates the equipment? _____

# Heat Pump (Heating Mode) Worksheet

Introduction: The service technician is often called upon to perform a capacity check of a heat pump system. This request may come in a variety of forms, such as a complaint of a high electric bill or not enough heat. Determining the capacity of a heat pump in the heating mode is not very difficult, and every service technician should perform this test whenever there is a suspected problem with the unit.

Tools Needed: dry bulb thermometer, velometer, gauge manifold, voltmeter, wattmeter, and tool kit.

Procedures:

1. Set the room thermostat to the heating or automatic position.

2. Set the temperature lever to demand heating.

3. Turn off all electricity to the auxiliary heat strips. Only the heat pump and the indoor fan are to be operating during this test.

4. Measure the temperature rise of the air through the indoor unit. Use the following formula:

    $$\Delta T = \text{Leaving air temperature} - \text{Entering air temperature}$$

    a. Use the same thermometer for measuring both the return and supply air temperatures.

    b. Do not measure the temperature "in view" of the indoor coil. True air temperature cannot be measured in areas affected by radiant heat.

    c. Make the temperature measurements within 6 ft of the indoor unit. Measurements taken at the supply and return air grilles are not accurate enough.

    d. Use the average temperature when more than one duct is connected to the plenum. Use the following formula:

    $$\text{Average} = \frac{\text{Total of readings}}{\text{Number of readings}}$$

    e. Make sure the air temperature is stable before taking these measurements.

    f. Take these measurements downstream from any mixed air source.

    g. Record the temperature difference of the return and supply air as $\Delta T$.

    $$\Delta T = \underline{\qquad} - \underline{\qquad}$$

5. Determine the Btu output. Use the following formula:

$$Btu = cfm \times 1.08 \times \Delta T$$

where: $cfm = \dfrac{Btuh}{1.08 \times \Delta T}$

$1.08$ = specific heat of air constant

$\Delta T$ = supply air temperature minus return air temperature

6. Is this what the manufacturer rates the equipment? _____

# Heat Pump (Heating Mode) Worksheet

Introduction: The service technician is often called upon to perform a capacity check of a heat pump system. This request may come in a variety of forms, such as a complaint of a high electric bill or not enough heat. Determining the capacity of a heat pump in the heating mode is not very difficult, and every service technician should perform this test whenever there is a suspected problem with the unit.

Tools Needed: dry bulb thermometer, velometer, gauge manifold, voltmeter, wattmeter, and tool kit.

Procedures:

1. Set the room thermostat to the heating or automatic position.

2. Set the temperature lever to demand heating.

3. Turn off all electricity to the auxiliary heat strips. Only the heat pump and the indoor fan are to be operating during this test.

4. Measure the temperature rise of the air through the indoor unit. Use the following formula:

    $$\Delta T = \text{Leaving air temperature - Entering air temperature}$$

    a. Use the same thermometer for measuring both the return and supply air temperatures.

    b. Do not measure the temperature "in view" of the indoor coil. True air temperature cannot be measured in areas affected by radiant heat.

    c. Make the temperature measurements within 6 ft of the indoor unit. Measurements taken at the supply and return air grilles are not accurate enough.

    d. Use the average temperature when more than one duct is connected to the plenum. Use the following formula:

    $$\text{Average} = \frac{\text{Total of readings}}{\text{Number of readings}}$$

    e. Make sure the air temperature is stable before taking these measurements.

    f. Take these measurements downstream from any mixed air source.

    g. Record the temperature difference of the return and supply air as $\Delta T$.

    $$\Delta T = \underline{\qquad} - \underline{\qquad}$$

5. Determine the Btu output. Use the following formula:

$$Btu = cfm \times 1.08 \times \Delta T$$

where: $cfm = \dfrac{Btuh}{1.08 \times \Delta T}$

$1.08$ = specific heat of air constant

$\Delta T$ = supply air temperature minus return air temperature

6. Is this what the manufacturer rates the equipment? _____

# Heat Pump (Heating Mode) Worksheet

Introduction: The service technician is often called upon to perform a capacity check of a heat pump system. This request may come in a variety of forms, such as a complaint of a high electric bill or not enough heat. Determining the capacity of a heat pump in the heating mode is not very difficult, and every service technician should perform this test whenever there is a suspected problem with the unit.

Tools Needed: dry bulb thermometer, velometer, gauge manifold, voltmeter, wattmeter, and tool kit.

Procedures:

1. Set the room thermostat to the heating or automatic position.

2. Set the temperature lever to demand heating.

3. Turn off all electricity to the auxiliary heat strips. Only the heat pump and the indoor fan are to be operating during this test.

4. Measure the temperature rise of the air through the indoor unit. Use the following formula:

    $\Delta T$ = Leaving air temperature - Entering air temperature

    a. Use the same thermometer for measuring both the return and supply air temperatures.

    b. Do not measure the temperature "in view" of the indoor coil. True air temperature cannot be measured in areas affected by radiant heat.

    c. Make the temperature measurements within 6 ft of the indoor unit. Measurements taken at the supply and return air grilles are not accurate enough.

    d. Use the average temperature when more than one duct is connected to the plenum. Use the following formula:

    $$\text{Average} = \frac{\text{Total of readings}}{\text{Number of readings}}$$

    e. Make sure the air temperature is stable before taking these measurements.

    f. Take these measurements downstream from any mixed air source.

    g. Record the temperature difference of the return and supply air as $\Delta T$.

    $\Delta T$ = _____ - _____

Copyright © Business News Publishing Company

5. Determine the Btu output. Use the following formula:

$$Btu = cfm \times 1.08 \times \Delta T$$

where: $cfm = \dfrac{Btuh}{1.08 \times \Delta T}$

$1.08$ = specific heat of air constant

$\Delta T$ = supply air temperature minus return air temperature

6. Is this what the manufacturer rates the equipment? _____

# Heat Pump (Heating Mode) Worksheet

Introduction: The service technician is often called upon to perform a capacity check of a heat pump system. This request may come in a variety of forms, such as a complaint of a high electric bill or not enough heat. Determining the capacity of a heat pump in the heating mode is not very difficult, and every service technician should perform this test whenever there is a suspected problem with the unit.

Tools Needed: dry bulb thermometer, velometer, gauge manifold, voltmeter, wattmeter, and tool kit.

Procedures:

1. Set the room thermostat to the heating or automatic position.

2. Set the temperature lever to demand heating.

3. Turn off all electricity to the auxiliary heat strips. Only the heat pump and the indoor fan are to be operating during this test.

4. Measure the temperature rise of the air through the indoor unit. Use the following formula:

    $\Delta T$ = Leaving air temperature - Entering air temperature

    a. Use the same thermometer for measuring both the return and supply air temperatures.

    b. Do not measure the temperature "in view" of the indoor coil. True air temperature cannot be measured in areas affected by radiant heat.

    c. Make the temperature measurements within 6 ft of the indoor unit. Measurements taken at the supply and return air grilles are not accurate enough.

    d. Use the average temperature when more than one duct is connected to the plenum. Use the following formula:

    $$\text{Average} = \frac{\text{Total of readings}}{\text{Number of readings}}$$

    e. Make sure the air temperature is stable before taking these measurements.

    f. Take these measurements downstream from any mixed air source.

    g. Record the temperature difference of the return and supply air as $\Delta T$.

    $\Delta T$ = _____ - _____

5. Determine the Btu output. Use the following formula:

$$Btu = cfm \times 1.08 \times \Delta T$$

where: $cfm = \dfrac{Btuh}{1.08 \times \Delta T}$

1.08 = specific heat of air constant

$\Delta T$ = supply air temperature minus return air temperature

6. Is this what the manufacturer rates the equipment? _____

# Heat Pump (Heating Mode) Worksheet

Introduction: The service technician is often called upon to perform a capacity check of a heat pump system. This request may come in a variety of forms, such as a complaint of a high electric bill or not enough heat. Determining the capacity of a heat pump in the heating mode is not very difficult, and every service technician should perform this test whenever there is a suspected problem with the unit.

Tools Needed: dry bulb thermometer, velometer, gauge manifold, voltmeter, wattmeter, and tool kit.

Procedures:

1.  Set the room thermostat to the heating or automatic position.

2.  Set the temperature lever to demand heating.

3.  Turn off all electricity to the auxiliary heat strips. Only the heat pump and the indoor fan are to be operating during this test.

4.  Measure the temperature rise of the air through the indoor unit. Use the following formula:

    $$\Delta T = \text{Leaving air temperature} - \text{Entering air temperature}$$

    a.  Use the same thermometer for measuring both the return and supply air temperatures.

    b.  Do not measure the temperature "in view" of the indoor coil. True air temperature cannot be measured in areas affected by radiant heat.

    c.  Make the temperature measurements within 6 ft of the indoor unit. Measurements taken at the supply and return air grilles are not accurate enough.

    d.  Use the average temperature when more than one duct is connected to the plenum. Use the following formula:

    $$\text{Average} = \frac{\text{Total of readings}}{\text{Number of readings}}$$

    e.  Make sure the air temperature is stable before taking these measurements.

    f.  Take these measurements downstream from any mixed air source.

    g.  Record the temperature difference of the return and supply air as $\Delta T$.

    $$\Delta T = \underline{\phantom{XXXX}} - \underline{\phantom{XXXX}}$$

Copyright © Business News Publishing Company

5. Determine the Btu output. Use the following formula:

$$Btu = cfm \times 1.08 \times \Delta T$$

where: $cfm = \dfrac{Btuh}{1.08 \times \Delta T}$

1.08 = specific heat of air constant

$\Delta T$ = supply air temperature minus return air temperature

6. Is this what the manufacturer rates the equipment? _____

# Refrigeration Worksheet

Introduction: An efficiency test should be performed annually on every commercial refrigeration unit. If this procedure is followed, the equipment will operate better, use less electricity, and last longer. The technician who can perform an efficiency test properly will always be in demand. Use the following procedure along with the proper instruments to determine the efficiency of refrigeration systems.

Tools Needed: dry bulb thermometer, gauge manifold, voltmeter, ammeter, velometer, and tool kit.

Procedures:

Use the following procedure for air-cooled condensers:

1. Visually check the entire system for cleanliness, and ensure all components are in proper working condition.

2. Start the unit and allow it to operate until the system pressures and temperatures have stabilized.

3. What type of system is this (high, medium, or low temperature)? _____

4. Install the gauge manifold, and record the pressures.
   Suction _____ psig, Discharge _____ psig

5. Measure the condenser leaving air temperature (db) and record. _____ °F

6. Measure the condenser entering air temperature (db) and record. _____ °F

7. Determine the temperature rise of the condenser air. Use the following formula:

    ΔT = Leaving air temperature - Entering air temperature

    ΔT = _____ °F

8. Determine the condenser cfm. Use the following formula:

    cfm = Area x Velocity

    cfm = _____

9. Calculate the heat rejected by the condenser. Use the following formula:

    CHR = cfm x 1.08 x ΔT x 0.30

    CHR = _____ Btuh

Copyright © Business News Publishing Company

10. Is this what the manufacturer rates the equipment? _____

Use the following procedure for water-cooled condensers:

1. Visually check the entire system for cleanliness, and ensure all components are in proper working condition.

2. Start the unit and allow it to operate until the system pressures and temperatures have stabilized.

3. What type of system is this (high, medium, or low temperature)? _____

4. Install the gauge manifold, read the pressures, and record.
   Suction _____ psig, Discharge _____ psig

5. Measure the leaving condenser water temperature and record. _____ °F

6. Measure the entering condenser water temperature and record. _____ °F

7. Measure the water flow through the condenser and record. _____ gpm

8. Calculate the heat rejected by the condenser. Use the following formula:

$$CHR = gpm \times 500 \times \Delta T$$

$CHR = $ _____ Btuh

9. Determine the effective refrigerating capacity of the equipment. Use the following formula:

$$ER = \frac{CHR}{15,000}$$

$ER = $ _____ Btuh

10. Is this what the manufacturer rates the equipment? _____

Use the following procedure for ice makers:

1. Visually check the entire system for cleanliness, and ensure all components are in proper working condition.

2. Start the unit and allow it to operate until the first harvest cycle is complete, plus an additional hour.

3. Remove the first harvest of ice.

4. Measure the entering water temperature where it starts to flow over the plates and record. _____ °F

Copyright © Business News Publishing Company

5. Measure the temperature of the harvested ice and record. _____°F

6. Weigh the total ice harvested for one hour and record. _____ lb

7. Determine the unit efficiency. Use the following formula:

$$HR = [(\text{Entering water temp.} - 32°F)] + 144 + [(32°F - \text{Ice temp.})(0.5)]$$

$$\text{Efficiency} = \frac{\text{lb of ice}}{HR}$$

Efficiency = _____ %

8. Is this what the manufacturer rates the equipment? _____

Copyright © Business News Publishing Company

# Refrigeration Worksheet

Introduction: An efficiency test should be performed annually on every commercial refrigeration unit. If this procedure is followed, the equipment will operate better, use less electricity, and last longer. The technician who can perform an efficiency test properly will always be in demand. Use the following procedure along with the proper instruments to determine the efficiency of refrigeration systems.

Tools Needed: dry bulb thermometer, gauge manifold, voltmeter, ammeter, velometer, and tool kit.

Procedures:

Use the following procedure for air-cooled condensers:

1. Visually check the entire system for cleanliness, and ensure all components are in proper working condition.

2. Start the unit and allow it to operate until the system pressures and temperatures have stabilized.

3. What type of system is this (high, medium, or low temperature)? _____

4. Install the gauge manifold, and record the pressures.
   Suction _____ psig, Discharge _____ psig

5. Measure the condenser leaving air temperature (db) and record. _____ °F

6. Measure the condenser entering air temperature (db) and record. _____ °F

7. Determine the temperature rise of the condenser air. Use the following formula:

   $\Delta T$ = Leaving air temperature - Entering air temperature

   $\Delta T$ = _____ °F

8. Determine the condenser cfm. Use the following formula:

   cfm = Area x Velocity

   cfm = _____

9. Calculate the heat rejected by the condenser. Use the following formula:

   CHR = cfm x 1.08 x $\Delta T$ x 0.30

   CHR = _____ Btuh

10. Is this what the manufacturer rates the equipment? _____

Use the following procedure for water-cooled condensers:

1. Visually check the entire system for cleanliness, and ensure all components are in proper working condition.

2. Start the unit and allow it to operate until the system pressures and temperatures have stabilized.

3. What type of system is this (high, medium, or low temperature)? _____

4. Install the gauge manifold, read the pressures, and record.
   Suction _____ psig, Discharge _____ psig

5. Measure the leaving condenser water temperature and record. _____ °F

6. Measure the entering condenser water temperature and record. _____ °F

7. Measure the water flow through the condenser and record. _____ gpm

8. Calculate the heat rejected by the condenser. Use the following formula:

$$CHR = gpm \times 500 \times \Delta T$$

   CHR = _____ Btuh

9. Determine the effective refrigerating capacity of the equipment. Use the following formula:

$$ER = \frac{CHR}{15,000}$$

   ER = _____ Btuh

10. Is this what the manufacturer rates the equipment? _____

Use the following procedure for ice makers:

1. Visually check the entire system for cleanliness, and ensure all components are in proper working condition.

2. Start the unit and allow it to operate until the first harvest cycle is complete, plus an additional hour.

3. Remove the first harvest of ice.

4. Measure the entering water temperature where it starts to flow over the plates and record. _____ °F

5. Measure the temperature of the harvested ice and record. _____°F

6. Weigh the total ice harvested for one hour and record. _____ lb

7. Determine the unit efficiency. Use the following formula:

$$HR = [(\text{Entering water temp.} - 32°F)] + 144 + [(32°F - \text{Ice temp.})(0.5)]$$

$$\text{Efficiency} = \frac{\text{lb of ice}}{HR}$$

Efficiency = _____ %

8. Is this what the manufacturer rates the equipment? _____

# Refrigeration Worksheet

Introduction: An efficiency test should be performed annually on every commercial refrigeration unit. If this procedure is followed, the equipment will operate better, use less electricity, and last longer. The technician who can perform an efficiency test properly will always be in demand. Use the following procedure along with the proper instruments to determine the efficiency of refrigeration systems.

Tools Needed: dry bulb thermometer, gauge manifold, voltmeter, ammeter, velometer, and tool kit.

Procedures:

Use the following procedure for air-cooled condensers:

1. Visually check the entire system for cleanliness, and ensure all components are in proper working condition.

2. Start the unit and allow it to operate until the system pressures and temperatures have stabilized.

3. What type of system is this (high, medium, or low temperature)? _____

4. Install the gauge manifold, and record the pressures.
   Suction _____ psig, Discharge _____ psig

5. Measure the condenser leaving air temperature (db) and record. _____ °F

6. Measure the condenser entering air temperature (db) and record. _____ °F

7. Determine the temperature rise of the condenser air. Use the following formula:

   $\Delta T$ = Leaving air temperature - Entering air temperature

   $\Delta T$ = _____ °F

8. Determine the condenser cfm. Use the following formula:

   cfm = Area x Velocity

   cfm = _____

9. Calculate the heat rejected by the condenser. Use the following formula:

   CHR = cfm x 1.08 x $\Delta T$ x 0.30

   CHR = _____ Btuh

10. Is this what the manufacturer rates the equipment? _____

Use the following procedure for water-cooled condensers:

1. Visually check the entire system for cleanliness, and ensure all components are in proper working condition.

2. Start the unit and allow it to operate until the system pressures and temperatures have stabilized.

3. What type of system is this (high, medium, or low temperature)? _____

4. Install the gauge manifold, read the pressures, and record.
   Suction _____ psig, Discharge _____ psig

5. Measure the leaving condenser water temperature and record. _____ °F

6. Measure the entering condenser water temperature and record. _____ °F

7. Measure the water flow through the condenser and record. _____ gpm

8. Calculate the heat rejected by the condenser. Use the following formula:

$$CHR = gpm \times 500 \times \Delta T$$

   CHR = _____ Btuh

9. Determine the effective refrigerating capacity of the equipment. Use the following formula:

$$ER = \frac{CHR}{15,000}$$

   ER = _____ Btuh

10. Is this what the manufacturer rates the equipment? _____

Use the following procedure for ice makers:

1. Visually check the entire system for cleanliness, and ensure all components are in proper working condition.

2. Start the unit and allow it to operate until the first harvest cycle is complete, plus an additional hour.

3. Remove the first harvest of ice.

4. Measure the entering water temperature where it starts to flow over the plates and record. _____ °F

5. Measure the temperature of the harvested ice and record. _____°F

6. Weigh the total ice harvested for one hour and record. _____lb

7. Determine the unit efficiency. Use the following formula:

$$HR = [(\text{Entering water temp.} - 32°F)] + 144 + [(32°F - \text{Ice temp.})(0.5)]$$

$$\text{Efficiency} = \frac{\text{lb of ice}}{HR}$$

Efficiency = _____%

8. Is this what the manufacturer rates the equipment? _____

# Refrigeration Worksheet

Introduction: An efficiency test should be performed annually on every commercial refrigeration unit. If this procedure is followed, the equipment will operate better, use less electricity, and last longer. The technician who can perform an efficiency test properly will always be in demand. Use the following procedure along with the proper instruments to determine the efficiency of refrigeration systems.

Tools Needed: dry bulb thermometer, gauge manifold, voltmeter, ammeter, velometer, and tool kit.

Procedures:

Use the following procedure for air-cooled condensers:

1. Visually check the entire system for cleanliness, and ensure all components are in proper working condition.

2. Start the unit and allow it to operate until the system pressures and temperatures have stabilized.

3. What type of system is this (high, medium, or low temperature)? _____

4. Install the gauge manifold, and record the pressures.
   Suction _____ psig, Discharge _____ psig

5. Measure the condenser leaving air temperature (db) and record. _____ °F

6. Measure the condenser entering air temperature (db) and record. _____ °F

7. Determine the temperature rise of the condenser air. Use the following formula:

    $\Delta T$ = Leaving air temperature - Entering air temperature

    $\Delta T$ = _____ °F

8. Determine the condenser cfm. Use the following formula:

    cfm = Area x Velocity

    cfm = _____

9. Calculate the heat rejected by the condenser. Use the following formula:

    CHR = cfm x 1.08 x $\Delta T$ x 0.30

    CHR = _____ Btuh

Copyright © Business News Publishing Company

10. Is this what the manufacturer rates the equipment? _____

Use the following procedure for water-cooled condensers:

1. Visually check the entire system for cleanliness, and ensure all components are in proper working condition.

2. Start the unit and allow it to operate until the system pressures and temperatures have stabilized.

3. What type of system is this (high, medium, or low temperature)? _____

4. Install the gauge manifold, read the pressures, and record.
   Suction _____ psig, Discharge _____ psig

5. Measure the leaving condenser water temperature and record. _____ °F

6. Measure the entering condenser water temperature and record. _____ °F

7. Measure the water flow through the condenser and record. _____ gpm

8. Calculate the heat rejected by the condenser. Use the following formula:

$$CHR = gpm \times 500 \times \Delta T$$

   CHR = _____ Btuh

9. Determine the effective refrigerating capacity of the equipment. Use the following formula:

$$ER = \frac{CHR}{15,000}$$

   ER = _____ Btuh

10. Is this what the manufacturer rates the equipment? _____

Use the following procedure for ice makers:

1. Visually check the entire system for cleanliness, and ensure all components are in proper working condition.

2. Start the unit and allow it to operate until the first harvest cycle is complete, plus an additional hour.

3. Remove the first harvest of ice.

4. Measure the entering water temperature where it starts to flow over the plates and record. _____ °F

5. Measure the temperature of the harvested ice and record. _____°F

6. Weigh the total ice harvested for one hour and record. _____ lb

7. Determine the unit efficiency. Use the following formula:

$$HR = [(\text{Entering water temp.} - 32°F)] + 144 + [(32°F - \text{Ice temp.})(0.5)]$$

$$\text{Efficiency} = \frac{\text{lb of ice}}{HR}$$

Efficiency = _____ %

8. Is this what the manufacturer rates the equipment? _____

# Refrigeration Worksheet

Introduction: An efficiency test should be performed annually on every commercial refrigeration unit. If this procedure is followed, the equipment will operate better, use less electricity, and last longer. The technician who can perform an efficiency test properly will always be in demand. Use the following procedure along with the proper instruments to determine the efficiency of refrigeration systems.

Tools Needed: dry bulb thermometer, gauge manifold, voltmeter, ammeter, velometer, and tool kit.

Procedures:

Use the following procedure for air-cooled condensers:

1. Visually check the entire system for cleanliness, and ensure all components are in proper working condition.

2. Start the unit and allow it to operate until the system pressures and temperatures have stabilized.

3. What type of system is this (high, medium, or low temperature)? _____

4. Install the gauge manifold, and record the pressures.
   Suction _____ psig, Discharge _____ psig

5. Measure the condenser leaving air temperature (db) and record. _____ °F

6. Measure the condenser entering air temperature (db) and record. _____ °F

7. Determine the temperature rise of the condenser air. Use the following formula:

   $\Delta T$ = Leaving air temperature - Entering air temperature

   $\Delta T$ = _____ °F

8. Determine the condenser cfm. Use the following formula:

   cfm = Area x Velocity

   cfm = _____

9. Calculate the heat rejected by the condenser. Use the following formula:

   CHR = cfm x 1.08 x $\Delta T$ x 0.30

   CHR = _____ Btuh

Copyright © Business News Publishing Company

10. Is this what the manufacturer rates the equipment? _____

Use the following procedure for water-cooled condensers:

1. Visually check the entire system for cleanliness, and ensure all components are in proper working condition.

2. Start the unit and allow it to operate until the system pressures and temperatures have stabilized.

3. What type of system is this (high, medium, or low temperature)? _____

4. Install the gauge manifold, read the pressures, and record.
   Suction _____ psig, Discharge _____ psig

5. Measure the leaving condenser water temperature and record. _____ °F

6. Measure the entering condenser water temperature and record. _____ °F

7. Measure the water flow through the condenser and record. _____ gpm

8. Calculate the heat rejected by the condenser. Use the following formula:

$$CHR = gpm \times 500 \times \Delta T$$

   CHR = _____ Btuh

9. Determine the effective refrigerating capacity of the equipment. Use the following formula:

$$ER = \frac{CHR}{15,000}$$

   ER = _____ Btuh

10. Is this what the manufacturer rates the equipment? _____

Use the following procedure for ice makers:

1. Visually check the entire system for cleanliness, and ensure all components are in proper working condition.

2. Start the unit and allow it to operate until the first harvest cycle is complete, plus an additional hour.

3. Remove the first harvest of ice.

4. Measure the entering water temperature where it starts to flow over the plates and record. _____ °F

5. Measure the temperature of the harvested ice and record. _____ °F

6. Weigh the total ice harvested for one hour and record. _____ lb

7. Determine the unit efficiency. Use the following formula:

$$HR = [(\text{Entering water temp.} - 32°F)] + 144 + [(32°F - \text{Ice temp.})(0.5)]$$

$$\text{Efficiency} = \frac{\text{lb of ice}}{HR}$$

Efficiency = _____ %

8. Is this what the manufacturer rates the equipment? _____

# Refrigeration Worksheet

Introduction: An efficiency test should be performed annually on every commercial refrigeration unit. If this procedure is followed, the equipment will operate better, use less electricity, and last longer. The technician who can perform an efficiency test properly will always be in demand. Use the following procedure along with the proper instruments to determine the efficiency of refrigeration systems.

Tools Needed: dry bulb thermometer, gauge manifold, voltmeter, ammeter, velometer, and tool kit.

Procedures:

Use the following procedure for air-cooled condensers:

1. Visually check the entire system for cleanliness, and ensure all components are in proper working condition.

2. Start the unit and allow it to operate until the system pressures and temperatures have stabilized.

3. What type of system is this (high, medium, or low temperature)? _____

4. Install the gauge manifold, and record the pressures.
   Suction _____ psig, Discharge _____ psig

5. Measure the condenser leaving air temperature (db) and record. _____ °F

6. Measure the condenser entering air temperature (db) and record. _____ °F

7. Determine the temperature rise of the condenser air. Use the following formula:

   $$\Delta T = \text{Leaving air temperature} - \text{Entering air temperature}$$

   $\Delta T =$ _____ °F

8. Determine the condenser cfm. Use the following formula:

   $$\text{cfm} = \text{Area} \times \text{Velocity}$$

   cfm = _____

9. Calculate the heat rejected by the condenser. Use the following formula:

   $$\text{CHR} = \text{cfm} \times 1.08 \times \Delta T \times 0.30$$

   CHR = _____ Btuh

Copyright © Business News Publishing Company

10. Is this what the manufacturer rates the equipment? _____

Use the following procedure for water-cooled condensers:

1. Visually check the entire system for cleanliness, and ensure all components are in proper working condition.

2. Start the unit and allow it to operate until the system pressures and temperatures have stabilized.

3. What type of system is this (high, medium, or low temperature)? _____

4. Install the gauge manifold, read the pressures, and record.
   Suction _____ psig, Discharge _____ psig

5. Measure the leaving condenser water temperature and record. _____ °F

6. Measure the entering condenser water temperature and record. _____ °F

7. Measure the water flow through the condenser and record. _____ gpm

8. Calculate the heat rejected by the condenser. Use the following formula:

$$CHR = gpm \times 500 \times \Delta T$$

   CHR = _____ Btuh

9. Determine the effective refrigerating capacity of the equipment. Use the following formula:

$$ER = \frac{CHR}{15,000}$$

   ER = _____ Btuh

10. Is this what the manufacturer rates the equipment? _____

Use the following procedure for ice makers:

1. Visually check the entire system for cleanliness, and ensure all components are in proper working condition.

2. Start the unit and allow it to operate until the first harvest cycle is complete, plus an additional hour.

3. Remove the first harvest of ice.

4. Measure the entering water temperature where it starts to flow over the plates and record. _____ °F

5.  Measure the temperature of the harvested ice and record. _____°F

6.  Weigh the total ice harvested for one hour and record. _____lb

7.  Determine the unit efficiency. Use the following formula:

$$HR = [(\text{Entering water temp.} - 32°F)] + 144 + [(32°F - \text{Ice temp.})(0.5)]$$

$$\text{Efficiency} = \frac{\text{lb of ice}}{HR}$$

    Efficiency = _____%

8.  Is this what the manufacturer rates the equipment? _____

# Refrigeration Worksheet

Introduction: An efficiency test should be performed annually on every commercial refrigeration unit. If this procedure is followed, the equipment will operate better, use less electricity, and last longer. The technician who can perform an efficiency test properly will always be in demand. Use the following procedure along with the proper instruments to determine the efficiency of refrigeration systems.

Tools Needed: dry bulb thermometer, gauge manifold, voltmeter, ammeter, velometer, and tool kit.

Procedures:

Use the following procedure for air-cooled condensers:

1. Visually check the entire system for cleanliness, and ensure all components are in proper working condition.

2. Start the unit and allow it to operate until the system pressures and temperatures have stabilized.

3. What type of system is this (high, medium, or low temperature)? _____

4. Install the gauge manifold, and record the pressures.
   Suction _____ psig, Discharge _____ psig

5. Measure the condenser leaving air temperature (db) and record. _____ °F

6. Measure the condenser entering air temperature (db) and record. _____ °F

7. Determine the temperature rise of the condenser air. Use the following formula:

    $\Delta T$ = Leaving air temperature - Entering air temperature

    $\Delta T$ = _____ °F

8. Determine the condenser cfm. Use the following formula:

    cfm = Area x Velocity

    cfm = _____

9. Calculate the heat rejected by the condenser. Use the following formula:

    CHR = cfm x 1.08 x $\Delta T$ x 0.30

    CHR = _____ Btuh

Copyright © Business News Publishing Company

10. Is this what the manufacturer rates the equipment? _____

Use the following procedure for water-cooled condensers:

1. Visually check the entire system for cleanliness, and ensure all components are in proper working condition.

2. Start the unit and allow it to operate until the system pressures and temperatures have stabilized.

3. What type of system is this (high, medium, or low temperature)? _____

4. Install the gauge manifold, read the pressures, and record.
   Suction _____ psig, Discharge _____ psig

5. Measure the leaving condenser water temperature and record. _____ °F

6. Measure the entering condenser water temperature and record. _____ °F

7. Measure the water flow through the condenser and record. _____ gpm

8. Calculate the heat rejected by the condenser. Use the following formula:

$$CHR = gpm \times 500 \times \Delta T$$

   CHR = _____ Btuh

9. Determine the effective refrigerating capacity of the equipment. Use the following formula:

$$ER = \frac{CHR}{15,000}$$

   ER = _____ Btuh

10. Is this what the manufacturer rates the equipment? _____

Use the following procedure for ice makers:

1. Visually check the entire system for cleanliness, and ensure all components are in proper working condition.

2. Start the unit and allow it to operate until the first harvest cycle is complete, plus an additional hour.

3. Remove the first harvest of ice.

4. Measure the entering water temperature where it starts to flow over the plates and record. _____ °F

5. Measure the temperature of the harvested ice and record. _____°F

6. Weigh the total ice harvested for one hour and record. _____lb

7. Determine the unit efficiency. Use the following formula:

$$HR = [(\text{Entering water temp.} - 32°F)] + 144 + [(32°F - \text{Ice temp.})(0.5)]$$

$$\text{Efficiency} = \frac{\text{lb of ice}}{HR}$$

Efficiency = _____ %

8. Is this what the manufacturer rates the equipment? _____

# Refrigeration Worksheet

Introduction: An efficiency test should be performed annually on every commercial refrigeration unit. If this procedure is followed, the equipment will operate better, use less electricity, and last longer. The technician who can perform an efficiency test properly will always be in demand. Use the following procedure along with the proper instruments to determine the efficiency of refrigeration systems.

Tools Needed: dry bulb thermometer, gauge manifold, voltmeter, ammeter, velometer, and tool kit.

Procedures:

Use the following procedure for air-cooled condensers:

1. Visually check the entire system for cleanliness, and ensure all components are in proper working condition.

2. Start the unit and allow it to operate until the system pressures and temperatures have stabilized.

3. What type of system is this (high, medium, or low temperature)? _____

4. Install the gauge manifold, and record the pressures.
   Suction _____ psig, Discharge _____ psig

5. Measure the condenser leaving air temperature (db) and record. _____ °F

6. Measure the condenser entering air temperature (db) and record. _____ °F

7. Determine the temperature rise of the condenser air. Use the following formula:

    $\Delta T$ = Leaving air temperature - Entering air temperature

    $\Delta T$ = _____ °F

8. Determine the condenser cfm. Use the following formula:

    cfm = Area x Velocity

    cfm = _____

9. Calculate the heat rejected by the condenser. Use the following formula:

    CHR = cfm x 1.08 x $\Delta T$ x 0.30

    CHR = _____ Btuh

Copyright © Business News Publishing Company

10. Is this what the manufacturer rates the equipment? _____

Use the following procedure for water-cooled condensers:

1. Visually check the entire system for cleanliness, and ensure all components are in proper working condition.

2. Start the unit and allow it to operate until the system pressures and temperatures have stabilized.

3. What type of system is this (high, medium, or low temperature)? _____

4. Install the gauge manifold, read the pressures, and record.
   Suction _____ psig, Discharge _____ psig

5. Measure the leaving condenser water temperature and record. _____ °F

6. Measure the entering condenser water temperature and record. _____ °F

7. Measure the water flow through the condenser and record. _____ gpm

8. Calculate the heat rejected by the condenser. Use the following formula:

$$CHR = gpm \times 500 \times \Delta T$$

   CHR = _____ Btuh

9. Determine the effective refrigerating capacity of the equipment. Use the following formula:

$$ER = \frac{CHR}{15,000}$$

   ER = _____ Btuh

10. Is this what the manufacturer rates the equipment? _____

Use the following procedure for ice makers:

1. Visually check the entire system for cleanliness, and ensure all components are in proper working condition.

2. Start the unit and allow it to operate until the first harvest cycle is complete, plus an additional hour.

3. Remove the first harvest of ice.

4. Measure the entering water temperature where it starts to flow over the plates and record.
   _____ °F

5. Measure the temperature of the harvested ice and record. _____°F

6. Weigh the total ice harvested for one hour and record. _____ lb

7. Determine the unit efficiency. Use the following formula:

$$HR = [(\text{Entering water temp.} - 32°F)] + 144 + [(32°F - \text{Ice temp.})(0.5)]$$

$$\text{Efficiency} = \frac{\text{lb of ice}}{HR}$$

Efficiency = _____ %

8. Is this what the manufacturer rates the equipment? _____

# Refrigeration Worksheet

Introduction: An efficiency test should be performed annually on every commercial refrigeration unit. If this procedure is followed, the equipment will operate better, use less electricity, and last longer. The technician who can perform an efficiency test properly will always be in demand. Use the following procedure along with the proper instruments to determine the efficiency of refrigeration systems.

Tools Needed: dry bulb thermometer, gauge manifold, voltmeter, ammeter, velometer, and tool kit.

Procedures:

Use the following procedure for air-cooled condensers:

1. Visually check the entire system for cleanliness, and ensure all components are in proper working condition.

2. Start the unit and allow it to operate until the system pressures and temperatures have stabilized.

3. What type of system is this (high, medium, or low temperature)? _____

4. Install the gauge manifold, and record the pressures.
   Suction _____ psig, Discharge _____ psig

5. Measure the condenser leaving air temperature (db) and record. _____ °F

6. Measure the condenser entering air temperature (db) and record. _____ °F

7. Determine the temperature rise of the condenser air. Use the following formula:

    $\Delta T$ = Leaving air temperature - Entering air temperature

    $\Delta T$ = _____ °F

8. Determine the condenser cfm. Use the following formula:

    cfm = Area x Velocity

    cfm = _____

9. Calculate the heat rejected by the condenser. Use the following formula:

    CHR = cfm x 1.08 x $\Delta T$ x 0.30

    CHR = _____ Btuh

Copyright © Business News Publishing Company

10. Is this what the manufacturer rates the equipment? _____

Use the following procedure for water-cooled condensers:

1. Visually check the entire system for cleanliness, and ensure all components are in proper working condition.

2. Start the unit and allow it to operate until the system pressures and temperatures have stabilized.

3. What type of system is this (high, medium, or low temperature)? _____

4. Install the gauge manifold, read the pressures, and record.
   Suction _____ psig, Discharge _____ psig

5. Measure the leaving condenser water temperature and record. _____ °F

6. Measure the entering condenser water temperature and record. _____ °F

7. Measure the water flow through the condenser and record. _____ gpm

8. Calculate the heat rejected by the condenser. Use the following formula:

$$CHR = gpm \times 500 \times \Delta T$$

   CHR = _____ Btuh

9. Determine the effective refrigerating capacity of the equipment. Use the following formula:

$$ER = \frac{CHR}{15,000}$$

   ER = _____ Btuh

10. Is this what the manufacturer rates the equipment? _____

Use the following procedure for ice makers:

1. Visually check the entire system for cleanliness, and ensure all components are in proper working condition.

2. Start the unit and allow it to operate until the first harvest cycle is complete, plus an additional hour.

3. Remove the first harvest of ice.

4. Measure the entering water temperature where it starts to flow over the plates and record. _____ °F

5. Measure the temperature of the harvested ice and record. _____°F

6. Weigh the total ice harvested for one hour and record. _____lb

7. Determine the unit efficiency. Use the following formula:

$$HR = [(\text{Entering water temp.} - 32°F)] + 144 + [(32°F - \text{Ice temp.})(0.5)]$$

$$\text{Efficiency} = \frac{\text{lb of ice}}{HR}$$

Efficiency = _____%

8. Is this what the manufacturer rates the equipment? _____

# Refrigeration Worksheet

Introduction: An efficiency test should be performed annually on every commercial refrigeration unit. If this procedure is followed, the equipment will operate better, use less electricity, and last longer. The technician who can perform an efficiency test properly will always be in demand. Use the following procedure along with the proper instruments to determine the efficiency of refrigeration systems.

Tools Needed: dry bulb thermometer, gauge manifold, voltmeter, ammeter, velometer, and tool kit.

Procedures:

Use the following procedure for air-cooled condensers:

1. Visually check the entire system for cleanliness, and ensure all components are in proper working condition.

2. Start the unit and allow it to operate until the system pressures and temperatures have stabilized.

3. What type of system is this (high, medium, or low temperature)? _____

4. Install the gauge manifold, and record the pressures.
   Suction _____ psig, Discharge _____ psig

5. Measure the condenser leaving air temperature (db) and record. _____ °F

6. Measure the condenser entering air temperature (db) and record. _____ °F

7. Determine the temperature rise of the condenser air. Use the following formula:

   $\Delta T$ = Leaving air temperature - Entering air temperature

   $\Delta T$ = _____ °F

8. Determine the condenser cfm. Use the following formula:

   cfm = Area x Velocity

   cfm = _____

9. Calculate the heat rejected by the condenser. Use the following formula:

   CHR = cfm x 1.08 x $\Delta T$ x 0.30

   CHR = _____ Btuh

10. Is this what the manufacturer rates the equipment? _____

Use the following procedure for water-cooled condensers:

1. Visually check the entire system for cleanliness, and ensure all components are in proper working condition.

2. Start the unit and allow it to operate until the system pressures and temperatures have stabilized.

3. What type of system is this (high, medium, or low temperature)? _____

4. Install the gauge manifold, read the pressures, and record.
   Suction _____ psig, Discharge _____ psig

5. Measure the leaving condenser water temperature and record. _____ °F

6. Measure the entering condenser water temperature and record. _____ °F

7. Measure the water flow through the condenser and record. _____ gpm

8. Calculate the heat rejected by the condenser. Use the following formula:

$$CHR = gpm \times 500 \times \Delta T$$

   CHR = _____ Btuh

9. Determine the effective refrigerating capacity of the equipment. Use the following formula:

$$ER = \frac{CHR}{15,000}$$

   ER = _____ Btuh

10. Is this what the manufacturer rates the equipment? _____

Use the following procedure for ice makers:

1. Visually check the entire system for cleanliness, and ensure all components are in proper working condition.

2. Start the unit and allow it to operate until the first harvest cycle is complete, plus an additional hour.

3. Remove the first harvest of ice.

4. Measure the entering water temperature where it starts to flow over the plates and record. _____ °F

Copyright © Business News Publishing Company

5. Measure the temperature of the harvested ice and record. _____°F

6. Weigh the total ice harvested for one hour and record. _____ lb

7. Determine the unit efficiency. Use the following formula:

$$HR = [(\text{Entering water temp.} - 32°F)] + 144 + [(32°F - \text{Ice temp.})(0.5)]$$

$$\text{Efficiency} = \frac{\text{lb of ice}}{HR}$$

Efficiency = _____ %

8. Is this what the manufacturer rates the equipment? _____

# Refrigeration Worksheet

Introduction: An efficiency test should be performed annually on every commercial refrigeration unit. If this procedure is followed, the equipment will operate better, use less electricity, and last longer. The technician who can perform an efficiency test properly will always be in demand. Use the following procedure along with the proper instruments to determine the efficiency of refrigeration systems.

Tools Needed: dry bulb thermometer, gauge manifold, voltmeter, ammeter, velometer, and tool kit.

Procedures:

Use the following procedure for air-cooled condensers:

1. Visually check the entire system for cleanliness, and ensure all components are in proper working condition.

2. Start the unit and allow it to operate until the system pressures and temperatures have stabilized.

3. What type of system is this (high, medium, or low temperature)? _____

4. Install the gauge manifold, and record the pressures.
   Suction _____ psig, Discharge _____ psig

5. Measure the condenser leaving air temperature (db) and record. _____ °F

6. Measure the condenser entering air temperature (db) and record. _____ °F

7. Determine the temperature rise of the condenser air. Use the following formula:

    $\Delta T$ = Leaving air temperature - Entering air temperature

    $\Delta T$ = _____ °F

8. Determine the condenser cfm. Use the following formula:

    cfm = Area x Velocity

    cfm = _____

9. Calculate the heat rejected by the condenser. Use the following formula:

    CHR = cfm x 1.08 x $\Delta T$ x 0.30

    CHR = _____ Btuh

10. Is this what the manufacturer rates the equipment? _____

Use the following procedure for water-cooled condensers:

1. Visually check the entire system for cleanliness, and ensure all components are in proper working condition.

2. Start the unit and allow it to operate until the system pressures and temperatures have stabilized.

3. What type of system is this (high, medium, or low temperature)? _____

4. Install the gauge manifold, read the pressures, and record.
   Suction _____ psig, Discharge _____ psig

5. Measure the leaving condenser water temperature and record. _____ °F

6. Measure the entering condenser water temperature and record. _____ °F

7. Measure the water flow through the condenser and record. _____ gpm

8. Calculate the heat rejected by the condenser. Use the following formula:

$$CHR = gpm \times 500 \times \Delta T$$

   CHR = _____ Btuh

9. Determine the effective refrigerating capacity of the equipment. Use the following formula:

$$ER = \frac{CHR}{15,000}$$

   ER = _____ Btuh

10. Is this what the manufacturer rates the equipment? _____

Use the following procedure for ice makers:

1. Visually check the entire system for cleanliness, and ensure all components are in proper working condition.

2. Start the unit and allow it to operate until the first harvest cycle is complete, plus an additional hour.

3. Remove the first harvest of ice.

4. Measure the entering water temperature where it starts to flow over the plates and record. _____ °F

Copyright © Business News Publishing Company

5. Measure the temperature of the harvested ice and record. _____°F

6. Weigh the total ice harvested for one hour and record. _____ lb

7. Determine the unit efficiency. Use the following formula:

$$HR = [(\text{Entering water temp.} - 32°F)] + 144 + [(32°F - \text{Ice temp.})(0.5)]$$

$$\text{Efficiency} = \frac{\text{lb of ice}}{HR}$$

Efficiency = _____ %

8. Is this what the manufacturer rates the equipment? _____

# Refrigeration Worksheet

Introduction: An efficiency test should be performed annually on every commercial refrigeration unit. If this procedure is followed, the equipment will operate better, use less electricity, and last longer. The technician who can perform an efficiency test properly will always be in demand. Use the following procedure along with the proper instruments to determine the efficiency of refrigeration systems.

Tools Needed: dry bulb thermometer, gauge manifold, voltmeter, ammeter, velometer, and tool kit.

Procedures:

Use the following procedure for air-cooled condensers:

1. Visually check the entire system for cleanliness, and ensure all components are in proper working condition.

2. Start the unit and allow it to operate until the system pressures and temperatures have stabilized.

3. What type of system is this (high, medium, or low temperature)? _____

4. Install the gauge manifold, and record the pressures.
   Suction _____ psig, Discharge _____ psig

5. Measure the condenser leaving air temperature (db) and record. _____ °F

6. Measure the condenser entering air temperature (db) and record. _____ °F

7. Determine the temperature rise of the condenser air. Use the following formula:

   $\Delta T$ = Leaving air temperature - Entering air temperature

   $\Delta T$ = _____ °F

8. Determine the condenser cfm. Use the following formula:

   cfm = Area x Velocity

   cfm = _____

9. Calculate the heat rejected by the condenser. Use the following formula:

   CHR = cfm x 1.08 x $\Delta T$ x 0.30

   CHR = _____ Btuh

Copyright © Business News Publishing Company

10. Is this what the manufacturer rates the equipment? _____

Use the following procedure for water-cooled condensers:

1. Visually check the entire system for cleanliness, and ensure all components are in proper working condition.

2. Start the unit and allow it to operate until the system pressures and temperatures have stabilized.

3. What type of system is this (high, medium, or low temperature)? _____

4. Install the gauge manifold, read the pressures, and record.
   Suction _____ psig, Discharge _____ psig

5. Measure the leaving condenser water temperature and record. _____ °F

6. Measure the entering condenser water temperature and record. _____ °F

7. Measure the water flow through the condenser and record. _____ gpm

8. Calculate the heat rejected by the condenser. Use the following formula:

$$CHR = gpm \times 500 \times \Delta T$$

$CHR = $ _____ Btuh

9. Determine the effective refrigerating capacity of the equipment. Use the following formula:

$$ER = \frac{CHR}{15,000}$$

$ER = $ _____ Btuh

10. Is this what the manufacturer rates the equipment? _____

Use the following procedure for ice makers:

1. Visually check the entire system for cleanliness, and ensure all components are in proper working condition.

2. Start the unit and allow it to operate until the first harvest cycle is complete, plus an additional hour.

3. Remove the first harvest of ice.

4. Measure the entering water temperature where it starts to flow over the plates and record.
   _____ °F

Copyright © Business News Publishing Company

5. Measure the temperature of the harvested ice and record. _____ °F

6. Weigh the total ice harvested for one hour and record. _____ lb

7. Determine the unit efficiency. Use the following formula:

$$HR = [(\text{Entering water temp.} - 32°F)] + 144 + [(32°F - \text{Ice temp.})(0.5)]$$

$$\text{Efficiency} = \frac{\text{lb of ice}}{HR}$$

Efficiency = _____ %

8. Is this what the manufacturer rates the equipment? _____

# Refrigeration Worksheet

Introduction: An efficiency test should be performed annually on every commercial refrigeration unit. If this procedure is followed, the equipment will operate better, use less electricity, and last longer. The technician who can perform an efficiency test properly will always be in demand. Use the following procedure along with the proper instruments to determine the efficiency of refrigeration systems.

Tools Needed: dry bulb thermometer, gauge manifold, voltmeter, ammeter, velometer, and tool kit.

Procedures:

Use the following procedure for air-cooled condensers:

1. Visually check the entire system for cleanliness, and ensure all components are in proper working condition.

2. Start the unit and allow it to operate until the system pressures and temperatures have stabilized.

3. What type of system is this (high, medium, or low temperature)? _____

4. Install the gauge manifold, and record the pressures.
   Suction _____ psig, Discharge _____ psig

5. Measure the condenser leaving air temperature (db) and record. _____ °F

6. Measure the condenser entering air temperature (db) and record. _____ °F

7. Determine the temperature rise of the condenser air. Use the following formula:

   $\Delta T$ = Leaving air temperature - Entering air temperature

   $\Delta T$ = _____ °F

8. Determine the condenser cfm. Use the following formula:

   cfm = Area x Velocity

   cfm = _____

9. Calculate the heat rejected by the condenser. Use the following formula:

   CHR = cfm x 1.08 x $\Delta T$ x 0.30

   CHR = _____ Btuh

Copyright © Business News Publishing Company

10. Is this what the manufacturer rates the equipment? _____

Use the following procedure for water-cooled condensers:

1. Visually check the entire system for cleanliness, and ensure all components are in proper working condition.

2. Start the unit and allow it to operate until the system pressures and temperatures have stabilized.

3. What type of system is this (high, medium, or low temperature)? _____

4. Install the gauge manifold, read the pressures, and record.
   Suction _____ psig, Discharge _____ psig

5. Measure the leaving condenser water temperature and record. _____ °F

6. Measure the entering condenser water temperature and record. _____ °F

7. Measure the water flow through the condenser and record. _____ gpm

8. Calculate the heat rejected by the condenser. Use the following formula:

$$CHR = gpm \times 500 \times \Delta T$$

   CHR = _____ Btuh

9. Determine the effective refrigerating capacity of the equipment. Use the following formula:

$$ER = \frac{CHR}{15,000}$$

   ER = _____ Btuh

10. Is this what the manufacturer rates the equipment? _____

Use the following procedure for ice makers:

1. Visually check the entire system for cleanliness, and ensure all components are in proper working condition.

2. Start the unit and allow it to operate until the first harvest cycle is complete, plus an additional hour.

3. Remove the first harvest of ice.

4. Measure the entering water temperature where it starts to flow over the plates and record. _____ °F

5. Measure the temperature of the harvested ice and record. _____ °F

6. Weigh the total ice harvested for one hour and record. _____ lb

7. Determine the unit efficiency. Use the following formula:

$$HR = [(\text{Entering water temp.} - 32°F)] + 144 + [(32°F - \text{Ice temp.})(0.5)]$$

$$\text{Efficiency} = \frac{\text{lb of ice}}{HR}$$

Efficiency = _____ %

8. Is this what the manufacturer rates the equipment? _____

# Refrigeration Worksheet

Introduction: An efficiency test should be performed annually on every commercial refrigeration unit. If this procedure is followed, the equipment will operate better, use less electricity, and last longer. The technician who can perform an efficiency test properly will always be in demand. Use the following procedure along with the proper instruments to determine the efficiency of refrigeration systems.

Tools Needed: dry bulb thermometer, gauge manifold, voltmeter, ammeter, velometer, and tool kit.

Procedures:

Use the following procedure for air-cooled condensers:

1. Visually check the entire system for cleanliness, and ensure all components are in proper working condition.

2. Start the unit and allow it to operate until the system pressures and temperatures have stabilized.

3. What type of system is this (high, medium, or low temperature)? _____

4. Install the gauge manifold, and record the pressures.
   Suction _____ psig, Discharge _____ psig

5. Measure the condenser leaving air temperature (db) and record. _____ °F

6. Measure the condenser entering air temperature (db) and record. _____ °F

7. Determine the temperature rise of the condenser air. Use the following formula:

   $\Delta T$ = Leaving air temperature - Entering air temperature

   $\Delta T$ = _____ °F

8. Determine the condenser cfm. Use the following formula:

   cfm = Area x Velocity

   cfm = _____

9. Calculate the heat rejected by the condenser. Use the following formula:

   CHR = cfm x 1.08 x $\Delta T$ x 0.30

   CHR = _____ Btuh

Copyright © Business News Publishing Company

10. Is this what the manufacturer rates the equipment? _____

Use the following procedure for water-cooled condensers:

1. Visually check the entire system for cleanliness, and ensure all components are in proper working condition.

2. Start the unit and allow it to operate until the system pressures and temperatures have stabilized.

3. What type of system is this (high, medium, or low temperature)? _____

4. Install the gauge manifold, read the pressures, and record.
   Suction _____ psig, Discharge _____ psig

5. Measure the leaving condenser water temperature and record. _____ °F

6. Measure the entering condenser water temperature and record. _____ °F

7. Measure the water flow through the condenser and record. _____ gpm

8. Calculate the heat rejected by the condenser. Use the following formula:

$$CHR = gpm \times 500 \times \Delta T$$

   CHR = _____ Btuh

9. Determine the effective refrigerating capacity of the equipment. Use the following formula:

$$ER = \frac{CHR}{15,000}$$

   ER = _____ Btuh

10. Is this what the manufacturer rates the equipment? _____

Use the following procedure for ice makers:

1. Visually check the entire system for cleanliness, and ensure all components are in proper working condition.

2. Start the unit and allow it to operate until the first harvest cycle is complete, plus an additional hour.

3. Remove the first harvest of ice.

4. Measure the entering water temperature where it starts to flow over the plates and record. _____ °F

5. Measure the temperature of the harvested ice and record. \_\_\_\_\_°F

6. Weigh the total ice harvested for one hour and record. _____lb

7. Determine the unit efficiency. Use the following formula:

$$HR = [(\text{Entering water temp.} - 32°F)] + 144 + [(32°F - \text{Ice temp.})(0.5)]$$

$$\text{Efficiency} = \frac{\text{lb of ice}}{HR}$$

Efficiency = _____%

8. Is this what the manufacturer rates the equipment? _____

# Refrigeration Worksheet

Introduction: An efficiency test should be performed annually on every commercial refrigeration unit. If this procedure is followed, the equipment will operate better, use less electricity, and last longer. The technician who can perform an efficiency test properly will always be in demand. Use the following procedure along with the proper instruments to determine the efficiency of refrigeration systems.

Tools Needed: dry bulb thermometer, gauge manifold, voltmeter, ammeter, velometer, and tool kit.

Procedures:

Use the following procedure for air-cooled condensers:

1. Visually check the entire system for cleanliness, and ensure all components are in proper working condition.

2. Start the unit and allow it to operate until the system pressures and temperatures have stabilized.

3. What type of system is this (high, medium, or low temperature)? _____

4. Install the gauge manifold, and record the pressures.
   Suction _____ psig, Discharge _____ psig

5. Measure the condenser leaving air temperature (db) and record. _____ °F

6. Measure the condenser entering air temperature (db) and record. _____ °F

7. Determine the temperature rise of the condenser air. Use the following formula:

    $\Delta T$ = Leaving air temperature - Entering air temperature

    $\Delta T$ = _____ °F

8. Determine the condenser cfm. Use the following formula:

    cfm = Area x Velocity

    cfm = _____

9. Calculate the heat rejected by the condenser. Use the following formula:

    CHR = cfm x 1.08 x $\Delta T$ x 0.30

    CHR = _____ Btuh

Copyright © Business News Publishing Company

10. Is this what the manufacturer rates the equipment? _____

Use the following procedure for water-cooled condensers:

1. Visually check the entire system for cleanliness, and ensure all components are in proper working condition.

2. Start the unit and allow it to operate until the system pressures and temperatures have stabilized.

3. What type of system is this (high, medium, or low temperature)? _____

4. Install the gauge manifold, read the pressures, and record.
   Suction _____ psig, Discharge _____ psig

5. Measure the leaving condenser water temperature and record. _____ °F

6. Measure the entering condenser water temperature and record. _____ °F

7. Measure the water flow through the condenser and record. _____ gpm

8. Calculate the heat rejected by the condenser. Use the following formula:

$$CHR = gpm \times 500 \times \Delta T$$

   CHR = _____ Btuh

9. Determine the effective refrigerating capacity of the equipment. Use the following formula:

$$ER = \frac{CHR}{15,000}$$

   ER = _____ Btuh

10. Is this what the manufacturer rates the equipment? _____

Use the following procedure for ice makers:

1. Visually check the entire system for cleanliness, and ensure all components are in proper working condition.

2. Start the unit and allow it to operate until the first harvest cycle is complete, plus an additional hour.

3. Remove the first harvest of ice.

4. Measure the entering water temperature where it starts to flow over the plates and record. _____ °F

Copyright © Business News Publishing Company

5. Measure the temperature of the harvested ice and record. _____°F

6. Weigh the total ice harvested for one hour and record. _____ lb

7. Determine the unit efficiency. Use the following formula:

$$HR = [(\text{Entering water temp.} - 32°F)] + 144 + [(32°F - \text{Ice temp.})(0.5)]$$

$$\text{Efficiency} = \frac{\text{lb of ice}}{HR}$$

Efficiency = _____ %

8. Is this what the manufacturer rates the equipment? _____

## Other Titles Offered by BNP

The Challenge of Recycling Refrigerants
Stain Removal Guide for Stone
Hydronics Technology
Fine Tuning Air Conditioning Systems and Heat Pumps
Refrigeration Reference Notebook
A/C and Heating Reference Notebook
Practical Cleanroom Design, Revised Edition
Managing People in the Hvac/r Industry
Troubleshooting and Servicing Modern Air Conditioning and Refrigeration Systems
Fire Protection Systems
Piping Practice for Refrigeration Applications
Born to Build: A Parent's Guide to Academic Alternatives
Industrial Refrigeration
Industrial Refrigeration, Volume II
Refrigeration Fundamentals
Basic Refrigeration
Refrigeration Principles and Systems
A/C, Heating, and Refrigeration Dictionary
How to Close, Second Edition
Service Agreement Dynamics
Plumbing Technology
Technician's Guide to Certification
Power Technology
Refrigeration Licenses Unlimited, Second Edition
A/C Cutter's Ready Reference
How to Solve Your Refrigeration and A/C Service Problems
Blueprint Reading Made Easy
Starting in Heating and Air Conditioning Service
Legionnaires' Disease: Prevention & Control
Schematic Wiring, A Step-By-Step Guide
Indoor Air Quality in the Building Environment
Electronic HVAC Controls Simplified
4000 Q & A for Licensing Exams
Heat Pump Systems and Service
Ice Machine Service
Troubleshooting and Servicing A/C Equipment
Stationary Engineering
Medium and High Efficiency Gas Furnaces
Water Treatment Specification Manual, Second Edition
The Four R's: Recovery, Recycling, Reclaiming, Regulation
SI Units for the HVAC/R Professional
Servicing Commercial Refrigeration Systems

TO RECEIVE A FREE CATALOG, CALL
**1-800-837-1037**

**BNP**
**BUSINESS NEWS**
**PUBLISHING COMPANY**
Troy, Michigan
USA

*Another Quality Book from Business News Publishing...*

# The Challenge of Recycling Refrigerants

By: Kenneth W. Manz

*The Challenge of Recycling Refrigerants* addresses the importance of proper refrigerant use, from installation to recovery; explains the economic impact of the phaseout of CFC and HCFC refrigerants; and provides the information service personnel need to make improved service choices.

*The Challenge of Recycling Refrigerants* further focuses on cleaning refrigerants for reuse, keeping refrigeration systems clean, protecting the used refrigerant supply, and analyzing how recovery-recycling equipment works to accomplish a variety of functions, with an emphasis on equipment design and selection criteria.

200 pages • Hardcover

Topics include:
- Recycling and renewable resources
- Recovery equipment
- Mixed refrigerants
- Choosing a recovery-recycling unit
- Moisture
- Oils
- Acids
- Particulates
- Retrofitting

## Order today!
## 1-800-837-1037

**Technical Books**
**The books of choice.**

# Technical Book Choices Should Be Black & White

# With BNP They Are!

## Call 1-800-837-1037

*Another Quality Book from Business News Publishing...*

## Troubleshooting and Servicing Modern Air Conditioning and Refrigeration Systems

By: John Tomczyk

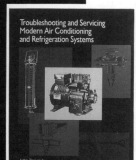

**Troubleshooting and Servicing Modern Air Conditioning and Refrigeration Systems** gives service technicians all the information needed to accurately diagnose and solve various system problems. This book includes information on the following topics:

- Refrigerant pressures, states, and conditions
- Subcooling and superheat
- Metering devices
- Compression systems
- Diagnosing air conditioning systems
- System charge
- Systematic troubleshooting
- Alternative refrigerants, refrigerant blends, and oils
- Leak detection, evacuation, and clean-up procedures

***Troubleshooting and Servicing Modern Air Conditioning and Refrigeration Systems*** emphasizes the changes affecting the refrigeration and air conditioning industries regarding CFC and HCFC refrigerants as well as alternative refrigerants and retrofit guidelines.

282 pages • Hardcover